Sound Works

Sound Works

A Cultural Theory of Sound Design

Holger Schulze

BLOOMSBURY ACADEMIC
NEW YORK • LONDON • OXFORD • NEW DELHI • SYDNEY

BLOOMSBURY ACADEMIC
Bloomsbury Publishing Inc
1385 Broadway, New York, NY 10018, USA
50 Bedford Square, London, WC1B 3DP, UK

BLOOMSBURY, BLOOMSBURY ACADEMIC and the Diana logo
are trademarks of Bloomsbury Publishing Plc

First published in the United States of America 2019
Reprinted 2019

Cover design by Louise Dugdale
Cover image © Julia Krause

Library of Congress Cataloging-in-Publication Data
Names: Schulze, Holger, author.
Title: Sound works : a cultural theory of sound design / Holger Schulze.
Description: New York, NY : Bloomsbury Academic, 2019. | Includes
bibliographical references and index.
Identifiers: LCCN 2019006883 (print) | LCCN 2019014045 (ebook) | ISBN
9781501330230 (ePDF) | ISBN 9781501330247 (ePub) | ISBN 9781501330223
(hardback)
Subjects: LCSH: Sound in design.
Classification: LCC TS171.4 (ebook) | LCC TS171.4 .S38 2019 (print) | DDC
004.01/9–dc23
LC record available at https://lccn.loc.gov/2019006883

A catalog record for this book is available from the Library of Congress.

ISBN: HB: 978-1-5013-3022-3
ePDF: 978-1-5013-3023-0
ePub: 978-1-5013-3024-7

Typeset by Integra Software Services Pvt. Ltd.

To find out more about our authors and books visit www.bloomsbury.com
and sign up for our newsletters.

CONTENTS

PREFACE

It was a Christmas morning, end of 2011. My wife, our three kids, and I were visiting my wife's parents at their cottage. The evening before, we had spoken at length about the research question and the methodology of the research project I was setting up at that time. It was a project on the cultural theory and the design theory of so-called *functional sounds:* the project behind this book (*cf.* the last section of Chapter 4: "The joy of research"). This morning, preparing breakfast for our siblings and relatives, we were suddenly alarmed by a cacophony of beeping and screeching noises: sitting at the large table we all realized that this cacophony was probably coming from some kitchen appliances. No doubt about that. However, none of us could discern—not even my wife's mother—what constellation of appliances was so desperately trying to alert us to something, to some completed task or to some urgent next step we would now need to take to finish our preparation of this breakfast. We burst out laughing.

Remembering this moment now—this funny and at the same time also irritating experience—it becomes clear how significant this random situation was for the state of functional sound in the early 2010s: a quite serendipitous demonstration of unresolved design issues. The cluster of minuscule beeping sounds, simultaneously routine and annoying, insistent and strangely timid, dominated all of a sudden a social situation. Most of the signal tones in this kitchen were so similar to each other that their genuine *sonic dominance* (Henriques 2011) pushed us immediately into utter disorientation. We stopped our tasks, we stared at each other, helplessly; reduced to clueless listeners, not knowing what functions all of these sounds should serve, what actions they should provoke right now. We just could not say which appliance was emitting this or that annoying signal? Was it the microwave oven? Was it the espresso machine? Was it the water heater? Or was it the large oven, or even just some smartphone? We had to laugh. A bodily and personal reaction to the irresolvable situational contradictions burst out of us; and obviously, while laughing, we also remembered our discussion the night before about precisely such functional sounds. The indiscernible cacophony of signaling was evident, sonic proof of all the issues with functional sounds these days.

Living with sound in contemporary networked and heavily mediated societies in the first quarter of the twenty-first century means living among sounds of various if not conflicting or incoherent origins. The domain of dealing with sound has been instrumental practice, composition, and—rather recently, historically speaking—musicology for decades and centuries now in the Western world; yet, most of the sounds you and I will actually encounter on any given day are definitely not arranged or invented to provoke radical aesthetic experiences or complex, autonomous reflections. These sounds of our everyday lives are strictly *functional.* They serve a certain ascribed purpose, they carry a heavy and very often unbearable agency if not an imperative, and they were crafted to transmit the intention of motivating or even ordering you or me to do or not to do certain things. They are products of sound design and communication design. But what does it mean to lead a life amongst functional sounds? How was it possible, historically, that such an ephemeral and transitory entity like *sound* could anyhow become the object of human design? And how are these sounds now actually designed by freelancers, by corporate communication departments, or in the course of developing and finalizing a product in the early twenty-first century? What if one leaves for once the perfect presentations of genius artist designers and brilliant sonic thinkers aside in order to focus on the often tedious everyday labor in one section of the creative industries caring for the sound layer of our societies? And what direct or long-term effects have the artifacts produced, implemented, and adjusted by this sonically creative industry on the everyday lives of you or me in the most intimate moments or the most professional contexts?

This study makes an effort to provide first tentative answers to these major questions. The four parts of this book follow distinct paths—and different readers might therefore find different parts more interesting. The first part of the book, *Why that Sound?*, provides a general introduction into the history, the theories as well as inherent aporias and working problems in the field of a design dealing with sound both as a source for excitement as well as annoyance; this part will be interesting for every reader from those who are new to sound design to experienced professionals. The second part, *Sonic Labor*, presents its argument in form of images, diagrams, and documents resulting out of our empirical study—participant observations, interviews, and questionnaires—with various sound designers. Statements by designers, characteristic work situations, and extracts of diverse case studies allow us to get a glimpse into the material if not trivial situation of actually designing sound; this part will be most attractive to sound designers, to journalists, and to researchers who like to get a quick insight into the everyday tasks, the material representations, and the current state of sound design. The third part, entitled *Living With Sound*, explores how these products of sound design processes are perceived, used or even misused by their intended users and consumers: how are sounds interpreted in the usage of particular situations of everyday life? This part takes its readers deep into the most intricate entanglements of discussions and current critiques in sound studies, in epistemology, in media studies, and in postcolonial studies based on our empirical studies on sound design and on additional field research. Existing approaches from semiotics are equipped here with a new and powerful engine as provided by the French cultural historian Régis Debray and his approach of *mediology*. This part therefore continues and deepens the discussions on sound design in the more specific fields of sound studies and media studies. The final and fourth part of this study, by the title of *Sound Works* proposes an experimental, at times provocative, and willfully extremist cultural theory of sound design: a theory that situates sound design in the midst of the force field of technology, politics, and society. This part connects our earlier observations, reflections, and insights in this book predominantly to the wider contemporary discussion in critical and in political theory, in social and in cultural theory. The insights presented in this final part allow for an understanding of sound design's role regarding society and politics in general. All in all, these four parts provide four research perspectives that are distinctively different from each other—yet obviously not unrelated but focusing on the same research object. As a result they grant an insight into the complexity and diversity of this very field of practice and its current interpretations.

At this early point of the book we also would like to warn our readers. If you expected this book to teach you how to design sound, being a kind of manual that lets you profit from the experience, knowledge, and various insider tricks as provided by world-class sound designers, then we have an urgent warning for you: *You need to stop reading this book – right now!* Against the background of such expectations this book can only bore you, annoy you, will disappoint you for sure. You will hate this book.

This book will *not* teach you sound design. This book, on the contrary, will let you explore the often quite disappointing and bleak labor conditions in sound design; it will allow you to explore how users of sound design live with functional sounds; in the end this book will propose an interpretation of the social, political, and anthropological effect sound design indeed has on everyday life and on contemporary cultures these days. This book is written for sound designers, for clients who commission a sound design, and for everyday users of sound design, and, obviously, for scholars of sound studies and related disciplines.

Holger Schulze

ACKNOWLEDGEMENTS

This book is conceived, written, and redacted in collaboration. This collaboration is not limited to the three authors whose names appear as collaborators to the chapters and parts of this publication. Artist and researcher Julia Krause was responsible for crucial parts of our empirical research on the everyday practices of sound designers as well as the illustrations documented in Part Two: "Sonic Labor." Sound studies and cultural studies scholar Carla J. Maier engaged in all the analytical research on everyday practices of listeners, the semiotics of sound, as well as the transcultural aspects of sound design documented in Part Three: "Living with Sound." I learned a lot from you both. A large number of designers, researchers, artists, and musicians have moreover contributed to the knowledge on sound design incorporated in this volume. Many of the conversations, collaborations, and field research-journeys had already started with conceiving the research project *Functional Sounds: Cultural Theory and Design Theory of Non-Verbal Sounds of Communication*, which was situated between 2011 and 2014, at the Humboldt-Universität Berlin and the Leuphana Universität Lüneburg with funding from the German National Research Foundation DFG. Preliminary research for this project began as early as 2001 on the masters program in sound studies and its various predecessors at the Berlin University of the Arts; post-processing interviews, additional field research, and surveys were executed from the Sound Studies Lab at the University of Copenhagen and on occasion of productions at the Deutschlandradio Kultur Berlin and at the Haus der Kulturen der Welt Berlin between 2014 and 2017. The authors wish to express their sincere thanks for all the insight and critique, all the productive complexifications and surprising encounters in production processes made possible by the following colleagues, collaborators, practitioners, and artists:

Peter Androsch, Sam Auinger, Leah Babb-Rosenfeldt, Nina Backmann, Stephen Barras, Karl Bartos, Ketan Batthi, Brandon LaBelle, Karin Bijsterveld, Mathias Black, Barry Blesser, Jochen Bonz, Karlheinz Brandenburg, Eoin Brazil, Kai Bronner, Hanna Buhl, Michael Bull, Zeynep Bulut, Thomas Burkhalter, Mark J. Butler, Joy H. Calico, Katherine De Chant, Caroline Churchill, Detlef Diederichsen, Diedrich Diederichsen, Melissa van Drie, Susanne Feld, Jens Fischer, Golo Föllmer, Nigel Frayne, Marcus Gammel, Klaus Genuit, Felix Gerloff, Heiner Goebbels, Steve Goodman, Antye Greie, Rolf Großmann, Sanne Krogh Groth, Wolfgang Hagen, Robert Heel, Florian Heesch, Edwin van der Heide, Thomas Hermann, Jan-Paul Herzer, Toby Heys, Elena Hidalgo, Milo Heller, Leopold Hoepner, Michiel Huijsman, Eleni Ikoniadou, Joseph Imorde, Detlev Ipsen, Anahid Kassabian, Ingo Kottkamp, Thomas Koch, Lutz Koepnick, Stefan Krebs, Jacob Kreutzfeld, Thomas Kusitzky, Yuri Landman, Christine Lang, Amy Martin, Kaffe Matthews, Christian Maier, Thomas Meinecke, Mathias Meppelink, Morten Michelsen, Guido Möbius, Marco Montiel-Soto, Satoshi Morita, Isabel Mundry, Adrian North, Bruce Odland, Dietmar Offenhuber, Pedro Oliviera, Norbert Ommer, Malte Pelleter, Katharina Pelosi, Markus Popp, Cornelius Ringe, Janine Rostrum aka Planningtorock, David Rothenberg, Daniel Salomon, Olaf Schäfer, Jens Scherzer, Matze Schmidt, Max Schneider, Brigitte Schulte-Fortkamp, Thomas Schwebel, Sebastian Schwesinger, Mark Smith, Ulrike Sowodniok, Georg Spehr, Erik Steinskog, Jonathan Sterne,

Hannes Strobl, Alfred Tesler, David Toop, Salomé Voegelin, Wolfgang Voigt, Urs Walter, Sebastian Waschulewski, Carl-Frank Westermann, Wessel Westerveld, Why Do Birds, Sascha Wild, Ariane Wilson, Justin Winkler, Stephan Winkler, Alexander Wodrich, Zeitblom, Renate Zentschnig.

Note: All illustrations and pictures are from the public domain—unless otherwise stated.

PART ONE

Why That Sound?

Annoyance and Excitement by Design

1

Working

Functional sounds in everyday life

Everyday sounds

When you turned on this computer—you are quite probably reading this book on—you had to listen to disjointed system sounds, start up chimes, alert noises, and auditory warning signals, perhaps accompanied by your favorite music in the background or on your headphones. Earlier today, when entering the building you work in, maybe you involuntarily set off the security alarm and the guards quickly materialized looking as if they would beat you up if you did not instantly produce your electronic identification card. Later, as you made a cup of coffee in the kitchen at work—or at home again—the automated (or humanly operated) coffee/espresso machine emanated a wide range of hissing, beeping, and crumbling noises. While drinking this pleasurable—and also quite functional—hot liquid you perhaps read a newspaper article about sound torture. And when you entered a subway station later on the same day, certain sounds announced almost every action taking place, escorted by sounds from various phones and game consoles. In bed at night you might have problems to fall asleep for all the annoying sounds coming from the traffic outside or the air conditioning system filling your rooms with its noise, hum, and resonance. This is everday life for you, for me, definitely not for all citizens on this planet, but for a lot.

Everyday life, these two supposedly tiny if not unimpressive words, imply a myriad of individual and idiosyncratic experiences in every single moment of every single hour on every single given day. An incoming message is being notified to me with a clicking signal sound by a social network I am logged into right now. Everyday life as such is unimaginable and hence it carries an unimaginable load of discourses, of academic discussions, of disagreement, and evermore diversifying approaches concerning the experience of the everyday (Pink 2012). So: What notion of the *everyday* is being used in this study? And on top, if this term really is of any use in academic discourse, what notion of *life* is being used here? In the back, the water heater signals with one distinct darker sound the end of the heating process. Implicitly, the concept of everyday life refers to its counter concept, the concept of the special and extraordinary, the rare, maybe even heterotopian moments in the lives of you or me. Special moments are what apparently many are seeking for, all the time, all their lives, also regarding sound design or of sound art. The special, extraordinary, and unusual of sound can be found in advanced high-end gadgets, in progressive concept studies, and highly elaborate performance settings for *Neue Musik* or *Klangkunst*. I tone down the volume as the streaming service sends its regular commercials. Such efforts to create auditory extravaganza manage with great joy and incredibly generous refinement to push the borders of what is technologically possible and what is desirable or digestible for potential listeners. These exciting developments of new sonic experiences and surprising aesthetics in sound they intend and succeed to educate you or me in listening as they represent the most radical outposts of sonic aesthetics: basic research in sound. Erratic subbass movements, subtle results and fusions of granular synthesis operations, unexpectedly joyful transitions of chords into melodies, put together out of fragmented layers of samples, glitches and clicks, oscillators and tone generators, mutually dependent and crafted in order to provide the most extremist, spatialized and at the same time haptic and kinesthetic sonic experience. Maybe a form of an *expanded site-specific rhythmanalysis* (Schulze 2018: 213)? Another social network I am logged into right now signals an incoming message with yet another, more melodic clicking sound.

At times, the trickle-down theory of sonic aesthetics can be observed: when new technological apparatuses and software inventions for crafting sounds begin as advanced developments of sound laboratories until they are finally assigned their successful and intensely monetarized place in the contemporary *economy of sound*—be it in popular music smash hits, in ubiquitous sound branding for kitchen appliances or all sorts of consumer electronics, in game sounds, the functional sounds of car navigation, or in the sounds of an amusement park. Yet, in most of the cases of newly developed auditory extravaganza this trickling down of aesthetics simply does not take place. Groundbreaking and deeply disturbing, erratic developments in sound do not actually dominate the lives of citizens in mediated societies en masse, twenty-four hours a day, seven days a week, fifty-two weeks a year, seventy-five years (or so) in one's life. I can hear the traffic sounds outside dying down; the traffic light slows them down. Therefore, the focus of this book on sound design does not lie on basic or advanced research in sound generation or composition; the focus of this book lies on the often rather mediocre, malfunctioning, or even horrible sounds, the sound signals, sound designs, and functional sounds that as a matter of fact do dominate your everyday and my everynight. In these moments and listening situations *everyday life* becomes a radical concept to focus on. It is the immanent situation most of the humanoid creatures on earth are enveloped in on every single moment in our lives.

The aforementioned *citizens* in heavily mediatized and networked societies can be considered a major type of listener to and user of sound designs, a type that matters. You swipe your card and you get granted access to your workplace. But if one normalizes this specific (or any certain) type of listener one surely ignores a wide array of other, maybe as important and as influential, examples of listeners that do not fit into this well-known and well-established pattern of a rather monolithic, a general concept of life as known to Western cultures: the concept of a male, white, well-educated, aristocratic-oligarchic, heterosexual, not bodily impaired lifeform, of employees and voters, of taxpayers and men and women living in heteronormative family institutions. I can hear myself typing, the acoustic feedback of my keyboard is just mechanic. To the contrary, this streamlined and idealized concept of life needs to be considered as thoroughly weird as it does nowhere on earth exist in exactly this normalized form. It is as such a pure imaginary, a representation of a social desire for normalization rather than a description of an actual lifeform on earth. Or in the words of the experimental psychologists Joseph Henrich, Steven J. Heine, and Ara Norenzayan:

> Western, Educated, Industrialized, Rich, and Democratic (WEIRD) societies [are] the least representative populations one could find for generalizing about humans. (Henrich, Heine, and Norenzayan 2010)

As soon as one abandons such an idealized and normative concept of "The Human" as this *WEIRD Human*, one is confronted with the surprising multiplicity of everyday lives in all their idiosyncratic richness, variety, and weirdness, between boredom and ecstasy, routine and surprise: potential listeners to these everyday sounds represent a multiplicity of erratic, plastic, and unforeseeably strange and alien ways of behaving, living, and appearing. This richness of the everyday, all the details of experiences and affects, of situations and plasticities, of interpretations and imaginations, of materializations and dissociations, is the foundation of this study in *radical empiricism* (James 1912). In the background, the hard drive, already some years old now, turns and makes its turning sounds audible. The material culture of designing sound and of living with designed sound is, hence, being explored in this study foremost in reference to specific, tiny, and ordinary situations of working on and living with sound. The traffic outside is running more loudly again. Starting with the everyday task of playing around with some rather arbitrarily generated sonic material, sketches and drafts for sound designs; the sudden urge to explicate a sound design idea in first tentative metaphors and phrases to a close colleague or a trustworthy project partner; the impromptu presentation of a first sketch in front of a group of selected people from the client in a rather unusual if not uncomfortable and inappropriate environment. On the ground of this specific and experiential reality and of its massive material impact the designed sound can be researched.

This radical focus on everyday situations of sonic labor leaves idealized listening situations and idealized concepts of genius designers way behind. It also leaves behind certain assumptions of ahistorically fixed meanings and culturally invariant messages in instrumental arrangements, in tonal progressions, or in rhythmic sequences. I can hear an old, 8-bit-ringtone in the distance. Static notions and systems of semiotics are suddenly rendered ahistoric and inappropriate—situated perception and activity, momentary impulses and accumulated training take control. Is this somehow the ancient Nokia-ringtone, cut, reduced, and designed from the *Gran Vals* by Francisco Tárrega written in 1902 (Gopinath 2013)? With the notion of everyday life and its *everyday sounds* this study refers to situations in individual lives that are not marked as special. In contrast such situations do often remain invisible, or one tends to forget for instance what exactly you or I did experience while taking the usual subway line on, for instance, that very Tuesday to our offices—or how one reacted as a colleague or a romantic partner or a good friend called while you were having lunch that one other Thursday—for example. The repeatability and the invisibility of everyday life experiences seem to make it incredibly hard to research in this field. Don't we prefer to remember all those strangely outstanding moments like *this one* weirdly unpleasant instance, or *that one* beautifully exciting encounter with a new acquaintance. All these extraordinary experiences though are the exact opposite to everyday life.

At this point the radical and rather unexpected impact of situations in everyday life becomes clearly audible: the gray and blurry background of one's everyday life is not faded out but put at the center of one's attention, on full volume, with almost unbearable loudness through the full bandwidth of all frequency bands. At the center of this study are *not* those nicely framing narrations of individual ambitions by sound designers in presenting the finest products of their efforts as the outcome of an always successful, never failing, always enjoyable, and never horribly devastating practice in a glorious way; this shall be the rightful domain of product presentations, of pitches and press releases to acquire new clients and customers. Instead, at the center of this study is the actual, the often humiliatingly banal, the dirty, and also confusing work environment and the not seldomly rather destructively erratic workflow in design projects involving sound design—as well as the erratic and strange, dissociated and unpredictable listening situations, the moments of misuse and misunderstanding, of disappropriation and of distortion in listening, in encountering, and in using sound design on a daily basis. I can hear the timer clock on the oven beeping from the kitchen.

All of this might sound rather disappointing, bleak, if not disenchanting for someone maybe hopeful to start a career in this promising and inspiring field of profession. But it is exactly in such bleak and disenchanted hours and days, working overtime, under working conditions of bricolage and imposture, where sound design actually is done. In these situations of disappointment, maybe boredom, surely repetition and routine the lives of producers and listeners of sound design unfold: in these many, these thousands and millions of hours, functional sounds occupy everyday life. I live among bleeps, you are punished by buzzers, we all are pinched and perforated by the affirming or denying sound signals of contemporary, sonified surveillance societies. This radical dystopia of contemporary social developments seems to be implied in the contemporary trajectories of sound design: for if one is restrained in this world of limited sonic experience then one is living under the rule of surveillance: a *Panacousticon* (*cf*. Chapter 14).

The dialectics of functionality

Sound designers get together in a workshop, on a regular basis. Doing a workshop has become part of developing a new sound design concept, reworking an existing or crafting a first tentative prototype for a sound design. One of the workshops I attended was aimed at developing a cross-platform and company-wide sound logo covering kitchen appliances, personal computer gadgets, mobile phone services, and even automotive navigation systems; and very soon, quicker than expected, the designers, musicians, accountants, and engineers engaged in a discussion that we could encounter in future workshops also: a discussion about the core criteria of an adequate, an aesthetically satisfying, and at the same time successful design. Very early, therefore, in our research it became quite clear that—in order to provide an appropriate foundation for our work on actual labor conditions and work processes in sound design—we would have to address also fundamental questions of design such as: What approaches to conceptualizing and designing sound could sound designers *potentially* rely on? Musical aesthetics? Sound art experiments? Usage studies? Psychoacoustics? The knowledge of auditory cognition? And, much more importantly: on what approaches to designing or conceptualizing sound do many of the sound designers we met *actually* rely on?

Compared to the fields of visual studies or of literary studies the field of *design research* is still rather in its nascent state (Mareis, Joost, and Kimpel 2010; Mareis, Held, and Joost 2013). However, a number of highly influential and prolific theories and publications can be identified that many designers rely on. In these approaches one can indeed discover generative elements for actual products of design in general. Be it the approach of *affordances* (Gibson 1977; Gaver 1991), of *design patterns* (Alexander 1977), of *design research* (Simon 1969; Laurel 2003; Cross 2007), or *design thinking* (Schön 1983; Sachse and Specker 1999; Lawson 2006; Plattner, Meinel, and Leifer 2010). All these approaches claim to provide tools for designers in order to structure, to reflect, to compare, and finally to improve their individual working processes. The process of designing is being commodified: articulation of a deep wish to make this process more controllable and more fitting to neighboring working processes, for example, in the sciences or in industry. In order to better understand the process of designing sound, it could hence be promising to not do the common but the unsuspected: not to expand only *sound theories* into sound design theories—but also to transfer *design theories* into sound design theories.

There are two intriguing and continuously influential approaches we will be focusing on in this study: the concept of a *pattern language* (Alexander 1977) and the concept of *affordances* (Gibson 1977; Gaver 1991). Starting with the first concept, the main questions of Christopher Alexander, mathematician and professor for architecture at Harvard University, in his design theory was: *How is it possible to design—originally in architecture—in a way that presents coherent, exhaustive and consequential solutions to general problems in design?* How can design leave some rather erratic efforts of imagination and trial and error behind and rely finally on empirically known examples from the past and their certain achievements concerning beauty and practicality in order to systematically present a so-called language for basic design solutions or design ideas? The 253 patterns proposed by Alexander include:

> 1. INDEPENDENT REGIONS ... 2. THE DISTRIBUTION OF TOWNS 3. CITY COUNTRY FINGERS 4. AGRICULTURAL VALLEYS 5. LACE OF COUNTRY STREETS ... 127. INTIMACY GRADIENT 128. INDOOR SUNLIGHT 129. COMMON AREAS AT THE HEART 130. ENTRANCE ROOM 131. THE FLOW THROUGH ROOMS 132. SHORT PASSAGES 133. STAIRCASE AS A STAGE 134. ZEN VIEW 135. TAPESTRY OF LIGHT AND DARK ... 249. ORNAMENT 250. WARM COLORS 251. DIFFERENT CHAIRS 252. POOLS OF LIGHT 253. THINGS FROM YOUR LIFE.

A remarkably exhaustive list. The patterns themselves are presented as a breathtakingly normative utopia of the thoroughly designed world: the whole earth shall be designed responsibly and thoughtfully. Pattern no. 130 "ENTRANCE ROOM," for instance, is outlined as:

> This pattern gives the entrances their detailed shape, their shape and body and three dimensions, and helps complete the form begun by car connection (113), and the private terrace on the street (140). (Alexander 1977: 623)

As in this example, the elements of design are presented as connected to each other and they seem to follow an intrinsically unraveling logic in the writings of Christopher Alexander. This approach, therefore, creates a language-like blueprint for design; needless to say: mainly in the limits of the hegemonic Western idea of a house and a habitat, though exemplified with cases from all over the world. But in these limits of a certain bourgeois and Western lifestyle the patterns by Alexander indulge in the joyfully detailed practical knowledge of everyday usage: "1. The relationship of windows to the entrance ... 2. The need for shelter outside the door ... 3. The subtleties of saying goodbye ... 4. Shelf near the entrance ... 5. Interior of the entrance room ... 6. Coats, shoes, children's bikes" (Alexander 1977: 623–625). The very minuscule practicalities of everyday life enter the design process. Take for instance Alexander's analytical narration of the "subtleties of saying goodbye," a truly moving and elucidating, almost ethnographical sketch. An applied cultural theory, full of practical knowledge and the humor of someone who enjoys the details of the everyday:

> When hosts and guests are saying goodbye, the lack of a clearly marked "goodbye" point can easily lead to endless "Well, we really must be going now," and then further conversations lingering on, over and over again. (a) Once they have finally decided to go, people try to leave without hesitation. (b) People try to make their goodbye as nonabrupt as possible and seek a comfortable break. Give the entrance room, therefore, a clearly defined area, at least 20 square feet, outside the front door, raised with a natural threshold—perhaps a railing, or a low wall, or a step—between it and the visitors' cars. (Alexander 1977: 624)

Here the experiential knowledge of everyday behavior, of anthropological inclinations and inhibitions, obsessions and fears becomes the foundation for inventing design solutions—for a certain Western lifestyle that assumes itself being cosmopolitan and global. This foundation in actual actions and desires of Western citizens connects Alexander's approach to the second approach mentioned earlier, James J. Gibson's concept of *affordance* (Gibson 1977) as a perceptual theory—especially in its expansion by William Gaver into a theory of *technological affordances* (Gaver 1991). This concept of affordance is originally a descriptive concept from the natural sciences, transferred into technology studies. Gibson summarizes the natural science approach of affordances as follows:

> The medium, substances, surfaces, objects, places, and other animals have affordances for a given animal. They offer benefits or injury, life or death. This is why they need to be perceived. The possibilities of the environment and the way of life of the animal go together inseparably. The environment constrains what the animal can do, and the concept of a niche in ecology reflects this fact. Within limits, the human animal can alter the affordances of the environment but is still the creature of his or her situation. (Gibson 1977: 143)

These general reflections on the anthropology of perception are applied on the design of devices and interfaces by William Gaver then in the following way:

> The notion of affordances is appealing in its direct approach towards the factors of perception and action that make interfaces easy to learn and use. As a means for analyzing technologies, affordances should be useful in exploring the psychological claims inherent in artifacts and the rationale of designs. More generally, considering affordances explicitly in design may help suggest ways to improve the usability of new artifacts. In providing an integrated account of a complex configuration of attributes, the concept provides a simple but powerful means of addressing a broad range of interface issues. ... It encourages us to consider devices, technologies and media in terms of the actions they make possible and obvious. It can guide us in designing artifacts which emphasize desired affordances and de-emphasize undesired ones. Perhaps most important, it allows us to focus not on technologies or users alone, but on the fundamental interactions between the two. (Gaver 1991: 5)

The anthropological affordances for designing actions as described by Alexander and the, if you will, more zoological patterns described by Gibson and developed by Gaver coherently form an implicit kind of *anthropology of design* that frames in their thinking a lot of contemporary design research. They also represent major dialectics inherent to the discourse among designers mentioned at the beginning of this section, regarding core criteria of an adequate, aesthetically satisfying, and successful design. Though the aspect of *functionality* rarely is addressed explicitly in the context of sound design, this issue lies apparently hidden underneath the issues concerning relevant approaches to conceptualizing and designing sound. On the one hand, to provide functionality in design—like Alexander proposes—pays respect to the actual, often rather intimate needs and desires, fears and obsessions of Western citizens living with design products; on the other side this focus on merely functional usage sequences and strict necessities and possibilities of action—like Gibson proposes—might lead in its design application to a narrowly, if not bossing guidance of a user who then becomes a merely constraining meatbag, an annoyingly erratic fleshsack in the acceleratingly natural process of a technologically dynamized society. Functionality, therefore, could be on the one side a result of basic anthropological respect or on the other side a simple way to dominate and control obedient citizens. Functionality has this double-faced quality in general—and for sound design in particular. The concept of functionality provides, not to forget, the ground for a long and winding history of theories of process optimization, of labor psychology, of fruitful reductionism, and of clever solutions to complex problems. A design theory for the twenty-first century, being also a result of sensory studies and sensorial practices, needs to navigate in the midst of this dialectics of functionality. Design cannot avoid any functionality; but design cannot restrict itself to provide only functionality. A *sound design theory,* understood as a division of design theories can therefore build upon this complex tradition of discussions and quarrels, doubts and ambitions. In our research on the actual labor in sound design we will try to provide a tentative approach to functionalities in sound design; and we will also propose concepts for sound designers to work with and for clients of sound designers to discuss present drafts: What could be the *sonic affordances* of a design assignment? And what kind of *sonic pattern language* would be developed in a sound design process (*cf.* Chapter 9, section “A sonic pattern language?” and Chapter 12, section “Sonic stereotyping”)?

The sonic consumer

Consumption is work. As soon as I approach the airport, as soon as I apparently intend to expose myself as being a consumer of aircraft tickets and airline products I am assimilated into this sensory designed consumption environment. I might feel tired, yet also relieved and light. The manifold sensory emanations of smell and sound, haptic and kinaesthetic sensation that represent this nearing experience of flying again are transmitted to me. Yet, this whole environment is a consumption environment. All of the sensory transmission I can assimilate are apparently intended to coerce me into consumption—or even to conceal the consumptory aspect of all my activities in this edifice. In the highly networked and globalized societies such a description at the beginning of the twenty-first century no longer qualifies as some form of leftist propaganda: it is recognized as an accurate and widely acknowledged description. In more recent approaches to understand the role of consumption in societies under the ideology of neoliberalism it becomes quite clear that the so-called *leisure time* is more and more structured like the so-called *time at work* (Fisher 2009). With a ping-sound another email is being dropped into your inbox. The important line once drawn between those two areas of life and of activity, between the hours in production and the hours of recreation to be fit for production, is even more blurred in the course of a so-called liberalization of restrictions for the workforce. Work is ubiquitous now. Consuming is ubiquitous. Recreation is ubiquitous. But why is it that suddenly all of these quite heterogeneous activities are experienced currently as a form of work, of plight, of obligation? With another clicking sound from the chat system of a social media network I am logged in, one project partner requests an immediate response to her earlier email. What necessary differentiation, what stratification or distinction in our societies and in our economies have been lost, have been blurred? Or were they always blurred but this blurring was simply ignored and neglected?

For research on the labor conditions in sound design this supposed transformation provides a quite unique situation: working in sound design, the product of this labor, this working *with* sound, constitutes at the same time also in part the conditions of this labor, a working *among* sounds. Working in sound design is a kind of *sonic labor* in this doubled definition. Therefore, the exploration of sound and its effects on listeners, employees, and citizens is not merely a side reflection, maybe added at the end of a long production process; already the production of a sound design is taking place under the conditions of contemporary sound design for various soft- and hardware products. Research on the anthropology of the senses and especially on the anthropology of sound is therefore not a fringe interest regarding sound design: it is directed at the main situations of production and of perception of sounds—in all the sound design projects that constitute the empirical foundation of this book. This doubled focus on sound will, as a consequence, accompany our research in the subsequent chapters of this book.

Sonic and sensory perforations of the everyday life have, this is safe to say in this early, introductory chapter, turned into an intense and massively capitalized project of an economy of mediatized and networked societies. What formerly had been media corporations, focusing on so-called *publications*, on so-called *transmissions, releases,* or *media packages* expanded consequentially into the whole realm of all things sensory: all the time, everywhere, under constant evaluation, reevaluation, and relaunch. The field of sensory experience once had been regarded as a field of liberty and freedom, of autonomous, aesthetic reflection, and of creation and consolidation of a civilized, modern subject. This once hopeful utopia of aesthetics has almost turned into its radical opposite. One of the most impactful insights of a historical anthropology of the senses is therefore that the senses are surely not any longer—and maybe they never were—an area of innocent, autonomous, and liberating self-exploration, self-expression, and liberation. In the twenty-first century the senses are commodified, reconceptualized, they are economically exploited to the max (*cf.* Howes 2005). Whereas in the nineteenth and twentieth centuries the most impactful forces on everyday life and politics came from major ideologies unraveled in elaborate treatises, volumes of studies and essays, this now has changed. It was the Italian philosopher Mario Perniola who coined for this transformation the term of *Sensology:* this term refers precisely to a shift from a scripture dominated ideology into new and contemporary forms of ideology one will not find in treatises—but embodied in gadgets, in technological apparatuses, in new cultural techniques and practices. The ideologies of present times are no longer mainly written down in large books, no longer mainly promoted in political rallies, on election day, or in partisan youth organizations—but they enter our lives as game consoles and VR-headsets, as swiping and tapping on touchscreens, as an ongoing sequence of tiny entertaining skits on streaming channels or television outlets, as a sequence of season finales and new season openers, of trailers, spoilers, of let's plays or walk-throughs. It is an ideology of and in and through the senses:

> In the present world, the circulation of sensologies has taken over the place of activity, the reflection and the echo of the already-felt have replaced thinking: acting and thinking are exposed to the continuous trade with sensological commodities, intending to sellout the entire contemporary universe. This means both the realization and the abolition of the metaphysical project; on the one hand, the externalization of feeling actually suppresses all autonomous impulses of the body and its affects, but on the other hand every primary intellectual effort is thwarted. (Perniola 1991: 134; translation by Holger Schulze)[1]

[1]"In der gegenwärtigen Welt hat die Zirkulation der Sensologien die Stelle der Tätigkeit übernommen, haben die Widerspiegelung und das Echo des Bereits-Gefühlten das Denken ersetzt: Handeln wie Denken sind dem ununterbrochenen Handel mit sensologischen Waren ausgesetzt, die es auf den Ausverkauf des gesamten zeitgenössischen Universums anlegen. Dies bedeutet sowohl die Umsetzung wie auch die Aufhebung des metaphysischen Projektes; einerseits ist durch die Entäußerung des Fühlens tatsächlich jeder autonome Impuls des Körpers und der Affekte unterdrückt, doch ist andererseits auch jedes Primat der intellektuellen Tätigkeit vereitelt" (Perniola 1991: 134).

This sensology turns any sensory awareness, any sensory activity and sensory life into a form of consumption. It has turned sensory existence into a form of commodification. You and I enjoy this—probably even at this very moment:

> There's no escaping. You are soaking in it. The adorables are coming for you. You will be mugged in the bright alleys of your dreams by the teddy bear horde, smothered in a vat of kitten memes, pounded with chiptune hooks until your ears bleed raspberry slurpie. Like the humidity of summer, cute will cling to you as you step out on the sidewalk, skin sticky and hot with the pressure of adorables flooding your senses, electric jingles echoing across the floor like loud orange bouncing balls saturating in pink-magenta halos, while sunlight that smells like gasoline pours through windows, and black coffee mingles with monoxide—oily and slick and strangely seductive with babyspeak chatter like unicorns on amphetamines vomiting rainbows. (Law and Wark 2013: 2f.)

We enjoy this, so much: our every nerve and sense for the desired, a hunch of fulfillment seems to be near—one tends to long for this. Who doesn't long for the fulfillment of desire? And this is so close, so quick, so simple, seemingly. The *sensory economy* of these days is in a certain sense the genuine heir to the media economy of the twentieth century. Though this media economy is still very present—if not selectively porous, questioned, crumbling, and not seldomly hollowed out—the new sensory economy is being established with a breathtaking speed. It spreads into all areas, every new season, every new quarter, every new month or week, brings new commodified versions to draw one's desire at—new gadgets, new series, new product lines, new platforms to create new accounts for, new profiles, new bonus badges to win, and new connections to make. The update to your preferred entertainment suite just dropped with a deep, dark bang into your download folder. Do you wish to install it right now? The sensory economy grants fulfillment 24/7. It is a joy, a true joy. One feels relieved and carried, safe and at home in consuming these sensory products, visual and kinaesthetic, these haptic and sonic products.

The sonic product, as grown out of the music industry and the communication agencies of recent decades, seduces one still with its invisibility and its ephemeral character, often a hunch of alchemy and the mysterious, if not otherworldly, expert knowledge, a secret knowledge mixed into a track, a sequence of sound files, a streaming platform. You turn it on and without anything actually changing in the material objects around you—well, aside from resonating vibrations, so to speak—everything changes around you, in you, with you:

> In everyday, mundane settings, it is most frequently through the ubiquity of musical sounds that we are reassured of the presence of something else, of more than ourselves: that's why people are in such a rush to put music on all the time, if it isn't already there. (Kassabian 2015: 557)

The perfect, sonic drug; the perfect sonic product. The subjectivities of listeners as consumers already are widely distributed to their preferred sonic service providers, streaming services, automated playlist accounts, aggregated preferences. This ongoing development concerning the artificial and commodified character of sensory events is, retrospectively, only foreshadowed in the massive commodification of sound events since the second half of the twentieth century. Jacques Attali stressed this prognostic quality of current transformations in sound cultures already decades ago (Attali 1977). The transformation of contemporary media market societies into collectives of sensory consumption is another example of this: We are *already* sensory consumers. We are *sonic consumers*. The perfect consumer citizens.

Functionality as guidance

Contemporary sound and consumer cultures under the spell of mediatized connectivity and an ongoing globalization and reordering along postcolonial frictions and scars, these cultures are populated more and more by functional sounds. As soon as you might leave this online repository for academic publications you might read these lines on, you will probably hear a certain confirming or rejecting sound; if this wasn't disabled or overruled by the music streaming or radio portal of your choice. *Consumer citizens* in highly connected cultures receive in everyday life their orders and signals of how to operate and where to focus, what to consume and in what ways to recreate by means of auditorily transmitted notifications. Whereas the concept of the consumer citizen is used in contemporary discourse (preferably by corporate think tanks and government commissions) as a concept to empower consumers to take their rights as citizens (Livingstone, Lunt, and Miller 2007), this concept will serve a more descriptive if not critical purpose in our research. Citizenship is in the refined cultures of communication, commodification, and consumption actually *only* existing as a form of consumption. Being a consumer seems, hence, to be an absolutely prerequisite for being granted citizen rights in the twenty-first century, such as free movement across borders, social security and healthcare, and even the status of being a juridical subject seems to be dependent on this quality. What is at this starting stage of the first chapter of this book more of an axiom, a claim of departure for our research, will be explored in greater detail and with a larger number of documents and examples in the following next two major parts of this book: *Sonic Labor* and *Living With Sound.* The wider impact of this dominant concept of a consumer citizen shall then be analyzed in the final sections of the last part of this book: *Sound Works.* At this point the departing axiom of a consumer citizen will only be unfolded tentatively, regarding their needs and desires toward sounds and sound designs.

In the early twenty-first century the globally hegemonic sound cultures are cultures of functionalized sound designs to order and to guide everyday lives. It is hence this desire for *guidance* that can be found at the heart of most of the designing efforts concerning sound. In contrast to any weirdly behaving, wild and untamed artifact, without clues for us as consumers or users any product with a certain functionality built in seemingly gives us hints and indices of how to use it and how to apply this product properly. Functionality makes us good users. Being a user though is not only characterized by a monodirectional agency: not only the user wants something from the artifact used—also this artifact requests certain actions, behavior, or even desire from the side of the user. In recent decades this mutual and recursive agency recurrently has been discussed under the metaphor of a mutual *domestication* of consumer and product. I follow here the quite complex and reciprocal definition as developed by Roger Silverstone and others in the 1990s and 2000s concerning the *cultural domestication of new media and technology.* Silverstone writes:

> The domestication of new media and information technologies involves, quite literally, a taming of the wild and a cultivation of the tame. (Silverstone 1994: 11)

In other words—translated into the field of sound culture and especially sound design products: *does the domestication of new products involve, quite literally, a taming of the wild and a cultivation of the tame by means of sound design?* Regarding sound design a so-called *wild* product—think of a nerdy, rough-cut prototype—is on the one side being tamed by the auditory notification signals that get to be attached to it and that shall help the consumers to understand, to use correctly, and not the least to like and love this wild beast. On the other side, though, these exact sound signals require themselves yet another quite different form of domestication, a *sonic domestication* to digest, to embody, and to finally appropriate the expectation and the adequate reaction to these sounds. *Sonic affordances* and *sonic patterns*—introduced as concepts in the earlier section on "The Dialectics of Functionality"—need to be learned and understood by every user anew. A focus on sound design as guidance and as a means to expand functionality is therefore not the final step in solving this issue of mutual domestication. Actually, it multiplies the problems as it elevates them to fulfill a much bigger purpose in the larger sensory economy, namely the domestication through guidance. Guidance in design products as a core element of the sensory economy is not only a semiotic, a haptic, or a performative issue but on top of all this a sensory, especially a sonic, issue.

Sensory guidance, consequentally, follows quite different and often neglected preferences and inclinations, cultural prerequisites and inhibitions as well as individual idiosyncrasies in usage, in abuse, and in nonuse. The case of sonic guidance in particular has its quite large pitfalls, and for outsiders quite surprising inner consistencies and requirements, that might be overseen from the perspective of product development and a general practice of product design in larger companies and corporations. These ambivalences and issues, these strange and unsuspected obstacles and genuine qualities in the process of designing sound are the main field of research for a cultural theory of sound design. A sound design—this is just one example from the sound design projects we researched and present in the following part on *Sonic Labor*—might, for instance, after a long process of development never actually be implemented, or one user might mute the whole sound design and another consumer might only recognize its aesthetic parallels to some other, earlier sound design—and only perceive it aesthetically and never actually use it. Functionality is, hence, easily overshadowed and overruled by other forces in usage. Still, it is the sound designers who are the major agents of reflecting, questioning, and also improving certain flaws in the institutionalized processes of designed sound. They are capable of doing so as they are at the same time also the first users and consumers of their own and others' functional sounds in the process of sonic labor. Sonic guidance is hence not a product that exists on some remote and strange islands outside of everyday life and work environments. Rather ironically the same humanoid aliens designing certain sounds, the sound designers, are at the same time their own sonic consumers: they are themselves suffering under earlier designed functional sounds of other sound designers—at their workplace, in their homes, whilst commuting.

In these situations, these moments, in which sound is created, in which sound designers work on specific, minuscule sound events, on a concept for sonic guidance for one requested product, or on a production technique to achieve a specific tone or timbre, in these moments all issues of sound design come together: one consumer citizen is working on a task, finding their way through the day, working on their own recreation in order to feel fresh, fit, and powerful enough to get back to work, and while doing so they are encountering sensory guidance, some sound events that intend to guide, to provoke, to nudge, to comfort or to order them in doing so. *The making of design and the usage of design are amalgamated in one moment.* Therefore this present exploration of a cultural theory of sound design necessarily has to acknowledge the double character of sound design: in order to unfold the details of our contemporary sound design culture it is necessary not only to research the individual ways and preferences of *Living With Sound*—as explored in the third part of this book—but it is necessary also to research in all its bleak humdrum of everyday work the individual ways and preferences of *Sonic Labor*—as explored in the subsequent part of this book, with reference to the seventeen distinct sound design projects that constitute the empirical foundation of this book. Sound designers are sound consumers, and sound consumers might design sound also. Both seek for sonic guidance by functional sounds, and both might at many points of their days be disturbed, be annoyed, be aggressively rejecting any sonic notification attacking them, perforating their hours, pinching their focus, their concentration, their individual concept of being a liberated being with a free will. The playlist playing whilst I am editing this section just stopped; a timid, gray signal sound occurred. *Functionality as guidance* can be a symptom of a paternalistic control society as well as it can be an expression of a thoughtfully caring craftsman. These dialectics of functionality will for sure be one watermark of this cultural theory of design. But before exploring the empirical details of all the conflicts at the workplace of designing *sonic pattern languages* and of living with *sonic affordances*, the introductory part of this book will continue to explore in the next three chapters the historical, the affective, and the desire-driven traits of experiencing and of inventing sound designs—the complexly layered cultural predispositions in our research for analyzing sound design projects as well as our lives with sound designs in the early twenty-first century.

2

Consuming

The brief history of a magic trick

The shock of sound

It comes as a shock. Suddenly, the sounds explode around you! In earlier periods of history—before the nineteenth century—the sounds coming from activities of fellow citizens, of animals, of slaves, of women, or of machinery were generally in mid-range loudness; only natural disasters, thunder, lightning, earthquake, or tsunami could shock you sonically by loudness, by volume, by its unexpected pressure waves. Anthropologically, most of the sounds around one's habitat, in chambers or villages, could be expected and foreseen. You could see, you could sense them coming—and if an *audio surprise* awaited you, indeed, even such a surprise would remain in mid-range, by standards of the early twenty-first century. This has radically changed since the notorious dawn of a new era with the various technologies of electric and magnetic transmission, of mediation, of storage, of amplification, and of reproduction. The advent of electrified and electronic media transformed the character, the impact, and the reach of sensory experiences as a whole. Since the appearance of recording, reproduction, and storage devices as well as the appearance of adapted theories, technologies, and usage of mediated transmission in all aspects of professional and personal life, the shock of sound recurrently plays a major role in everyday life. This shock of sound embodies a shock of media, a shock of the new—Toffler's ancient *Future Shock* would probably be sonified as a sonic boom, as a nuclear explosion (Toffler 1970). A *sonic dominance* (Henriques 2011) in historical perspective.

This shock of sound though is, essentially, just a magic trick. Most of the time you *assumed* having heard a mediated recording of a sonic boom—or a nuclear explosion, or a space ship accelerating into hyperdrive—it almost surely was *not* actually a sonic boom. Recording sonic booms is no simple task for standard equipment. Reproducing a sonic boom in all its frequency range and loudness is hence near to impossible, by definition: it would destroy the equipment—or at least it would exceed its range of recordable percepts. What you actually heard in all of those cases were mostly those pressure waves an ordinary standard equipment might be capable of recording; yet if you listened to a recorded sonic boom in a movie in all its brilliance and subtlety—it is safe to say—it was an artificially designed and post-produced sound event. *You did not hear* a sonic boom: you heard a sonically performed magic trick. Since then you believe this magic trick sound to be the actual, audible reality of a sonic boom.

The mediated phenomenon of *hyperrealism*, Jean Baudrillard wrote about on occasion of the visual extravaganza of the Hollywood movie industry, is of even greater impact regarding the sounding reality (Baudrillard 1994). *Sonic hyperrealism* is almost everyone's standard experience. You and I, we assume to know for sure how everyday events sound—just by their massively enhanced, post-produced, and polished-up media effects. Yet, what one remembers for instance as the sound of a space ship exploding in space, or the sound of a large sword bursting open the ribcage of a medieval fantasy warrior, or the sound of a cigarette being lighted, the tobacco gloriously going up in flames, subsequently burning up into ashes: all these examples represent—in Baudrillard's words—foremost "a real without origin or reality" (Baudrillard 1994: 1).

This history of sound design as a magic trick started with the first auditory gimmicks in the deep history of markets and fairgrounds; over the *trompe l'oreille* sounds of traditional theatre or in vaudeville, the seemingly natural noises produced by shaking the so-called *thunder sheet* or by working at a *rain-* or a *wind-machine*; up to the more recent, properly named sound design, following the major inventions by Jack Foley as well as the wide variety of spatialized sound effects in the most advanced image and sound projection productions for blockbuster movies—as conceptualized and executed for instance by the likes of Ben Burtt, Walter Murch, Gary Rydstrom, or Steve Tushar. These sound designers recreated, emulated, and actually shaped and crafted sonic experiences for generations. They invented those magic tricks that lead one to hear and to experience and to enjoy the materiality, the immediacy, and the complexity of any event as massively real, *hyperreal*. In 2001, the Swiss researcher Barbara Flückiger published one of the most impressive theoretical accounts of how sound design for movies is generated and how it could be conceptualized (Flückiger 2001). Her main focus points for the analysis of sound designs are, so to speak, a compressed, rather pragmatic, and very illuminating account of the framework of sonic materialism—systematically building upon some of the most pronounced theoretical approaches from decades ago by Pierre Schaeffer (1966), Murray Schafer ([1977] 1994), and Michel Chion (1990): *What Is Sounding? What is Moving? What Material is Sounding? How Does it Sound? Where Does it Sound?* (Flückiger 2001, 2009, *passim*). With these five simple questions Flückiger manages to hand over a tool that helps in analyzing what she calls the *USO*—the *Unidentified Sound Object*. Any product of sound design that follows the tradition of the magic trick, the intention to hyperrealize an action by sonifying it by means of several production, recording, and post-production techniques, can be called an *unidentified sound object*; as the sound might be, hopefully, identified—yet it remains, also hopefully, unclear how this sound event actually has been produced. One might be surprised that the sound you hear is not actually produced by the recording of a burning cigarette, a stabbed warrior, or an exploding space ship. Flückiger writes:

> The basic characteristic of unidentifed sound objects is that they have been severed from any connection to a source. A source is neither visible nor may it be inferred from the actual context. In addition, spectators are denied any recognition cues. In general the level of ambiguity is not reduced but carefully maintained in order to build up emotional tension in the viewer/listener. (Flückiger 2009: 156)

This is the magic trick of sound design. The trickery though is not anymore restricted to ghost trains or thunder in the theatre, to Foley sound or digital postproduction; in the twenty-first century the magic trick has expanded into almost the least niche of everyone's personal and professional life—in those mediated, and networked, and excessively commodified societies in which you and I are supposedly spending most of our days. What once was the reason for some extraordinary excitement, for outrageous experiences of kinaesthetic, multisensory extravaganza, rather extreme moments in one's life, completely out of the ordinary, these moments have now infiltrated the routines of everyday tasks, of daily routine. This sensory excess has turned from an unusual, often otherworldly, at times almost spiritual experience, into a rather ordinary exercise, an integral and indispensable trait of the daily lives of us consumer citizens. You and I—we both expect such subtle, pointed, and unthinkably refined sound productions even in the least mediated sound events one could imagine: in the beeping echolocation sound of an acoustic parking system in one's car's navigation, in the sound with which your kitchen appliance might signal a need for fill-up, for repair, or just for a regular check-up, in the notification sounds of some online service or platform, in the ringtone blasting out of our mobile communication device—just some examples of the sound design projects we encountered and followed in certain or all stages in their design process. Sound design is a magic trick having entered our everyday lives. A commodified sonic shock, an excitement gone blasé.

Magic mediation

The sensory shock of amplified (and often joyfully distorted) sound still lingers on. For a broader audience, as well as for an expert public, sound design research is apparently still one of the most attractive fields of research in sound studies. This positive projection, luckily, helped us a lot in the course of our research. Contrary to this public interest, there is even today still no comprehensive, historically and critically profound theory on sound design. The research recently done in this field (e.g., Spehr 2008, 2009; Pysiewicz 2009; Supper 2012; Gopinath 2013; Bijsterveld et al. 2014; Steiner 2014) and the still highly influential and illuminating publications by informed practitioners (e.g., Jackson 2004; Bronner 2004; Bronner and Hirt 2009, 2016; Treasure 2010; Groves 2011; Weiss 2015) are living proof of the general assumption that this field is still in the stage of an elaborate practice requiring highly sophisticated technical knowledge on the one hand, and a highly inspired and thus quite opaque aesthetic genius on the other. A field of *magic mediation*. To make this perfectly clear: one of the most ambitious and inspiring, technologically advanced and immersive fields of design practice relies still today *almost exclusively* on autodidactic skills, on personal ambition, and on an individual earnestness of its protagonists. In such a state, this field of design reminds one, frankly speaking, more of premodern concepts of an artistic genius and of a skilled witchcraft: concepts that harshly contradict and hence falsify all contemporary imagery of sound designers and sound engineers being thoroughly rational masters of earnest and down-to-earth signal processing, businesslike creators of computerized sonic emissions. If sound design indeed would be more a kind of mediated magic, how could it ever be evaluated, rationally conceptualized, and commerically marketed?

Karin Bijsterveld and her research group at the University of Maastricht recently confirmed this major issue for the field of sound design, judging from research on automotive sound design: *how to evaluate—historically, transculturally, and non-biased—the actual quality, the effect, and the impact of a sound design?* Especially in times when the highly sophisticated sound design of movies—as could be seen in the previous section—overtakes everyday lives, in times when all forms of appliances have rather individual sounds and require radically idiosyncratic forms of *domestication* (Silverstone 1994), and finally when immersion as desired in movies is almost impossible to achieve in everyday usage of sound designed products implemented into one's individual life: How could it be possible to measure actually in a convincing—and not bullshitting—way the effects and affects of sound design? Can this magic trickery be proven? Bijsterveld and her team found this:

> Just like the aircraft engineers, the acoustical engineers who aimed at concrete sound-design specifications for cars had to construct a "realistic" setting for *listening* to car sounds in which the input of sound could be measured precisely. Similar to the aircraft engineers, moreover, the acoustics people needed to find ways to have the test subjects *verbalize* their evaluation of sound in a maximally sophisticated manner. (Bijsterveld and Cleophas 2012: 109; emphasis in the original)

The evaluation of sound design hence encounters two obstacles: firstly, the situation of measuring would need to be reproducible to conduct a valid amount of measurements; secondly, the listening test subjects should be able to verbalize their affects—maybe it would be necessary to first educate them how to correctly verbalize a *sonic impression*? The limits of objectifying affects become quite obvious here—and with them also the limits of establishing sound design as a form of commodified research following the epistemic models of the engineering sciences, of experimental psychology, or of natural sciences. As the researchers state:

> Only after prolonged close cooperation between engineers, instrument makers, and pilots at a particular laboratory did the aircraft industry manage to develop a set of precise instruments and piloting opinion, so that the pilot "knew what he *liked* and could *say* so." (Vincenti 1990: 103, quoted in Bijsterveld and Cleophas 2012: 108; emphasis in the original)

So the pilots—notably *the main users* of these exclusively functional sounds of an aircraft—had first of all to learn to appreciate, to evaluate, and simply to judge the sensorial and aesthetic experience of the sounds. The magic of sound, far from being unquestionable and intuitively effective, hence has first to be mediated to become a reflected part of professional and public discourses and their cultural processes. Evaluating sound design has not been cultivated by us consumer citizens over a long period; it is a cultural practice whose refinement is still in process. The users of sound designed products therefore learn not only how to operate these products (here: an aircraft) but also how to evaluate, how to react, and how to adapt to a certain sound design in contrast to other sound designs they know. It is a professional, an engineering, and at the same time an aesthetic discourse that needs to be learned alongside the pragmatics. *The magic needs to be mediated. You need to domesticize the trick*. It is not just there. The way you might have learned to read visual accounts of statistics, learned to understand the meaning of traffic signs, or learned how to read a musical score or the visual representation of recorded tracks on the screen of a digital audio production application, is the same way you need to learn in the course of your life how certain sound signals are to be answered, how you are supposed to react, and how older functional sounds maybe have changed or retained their meanings with a new sound. As sonic consumers we adapt to the *sonic affordances* of these new sound signals—and we might recognize some of the *sonic pattern languages* acquired earlier in our lives. Design theories introduced at an earlier stage of sound design history must hence prove useful and appropriate for sound design at a later, future stage.

This process of a magic mediation, this transmission of the trick of the sonic boom is taking place in most of the publications by informed and experienced practitioners such as Julian Treasure, Kai Bronner, John Groves, or Peter Philipp Weiss. All of these publications provide an insight into the making of sound design that proved helpful in our encounters with the selected sound design projects that constitute the empirical basis of this book. But some of these do achieve this by discussing and partially including acknowledged strategies of research and also the recognition of various results from the scientific community (as Bronner 2004; Bronner and Hirt 2009, 2016; Weiss 2015)—and sometimes they present more of an individual, actually even artistic reflection by a designer or a team of designers on his (very rarely, still, *her*) personal approach to this field of design (as, for example, Jackson 2004, Treasure 2010, Groves 2011). In any case these writings present also numerous case studies and anecdotes, sometimes with a more self-critical, and sometimes with a more self-congratulatory if not shamanistic flavor. One of the most extreme examples of such a more, let's say, motivational and almost spiritual writing can probably be found in the first lines of the preface in Daniel Jackson's crucial and highly influential book by the name of *Sonic Branding*:

> Be yourself. Be honest and express how you feel. Listen to how others feel, empathize and try to give them what they want and need. Treat everybody in this way, every time you meet them. Make an effort and think it through. (Jackson 2004: xiii)

The magic mediation begins—right here. Be prepared for the wonders to come. Or—not?

Substantial disappointment

But what if these wonders never occur? What if the enlightened propagandists of sound design evoke an expectation of quality in sonic experiences in their audience that never can actually be met by actual sound design in everyday life? The branding sound of a railroad company—one of the projects we encountered—might not actually have the desired effect on its passengers, or the new functional sound of an endoscopic machine—another sound design project in our fieldwork—does indeed vanish in the multiplicity of other sounds. What if you turn, later on, to a praised sound design and you find it, well, rather underwhelming, unflattering, if not boring or just, all in all, sloppily sticked together? In the modern history of audiovisual media, besides the magic trick of announcing and often providing hitherto unknown sensory sensations, the other side of this magic trick, this thrilling promise of the sensational, is only rarely mentioned or theorized. Every new medium, every new genre, and every individual biography of experiencing new products provides—according to media history—not only the well-known accounts of utter excitement and flabbergasting insanity; these assumed new and shocking media apparatuses suffer another effect from the side of their recipients and users. By hearing, seeing, and reading in all the massive advertisements about the unimaginable wonders of sound and vision to be expected in this new, nameless thing, the future users of this new thing will undoubtedly develop a fantasy, an anticipating imagination of how unexpected and unusual and life changing the experience of this new thing might probably be. This imagination is fueled not only by the advertisements, the impressively convincing speeches and presentations from product marketers in globally streamed conventions, eagerly anticipated by future consumers, fanboys and fangirls of this brand. This imagination is also fueled by all the individual hopes, fears, the desires and the insecurities of all the consumer citizens expecting to buy and to domesticate and to use this new *thing* on a daily basis. But, no actual product and no existing invention will ever be able to meet these insanely bloated expectations, no screen, no virtual environment, no sound reproduction gadget will ever be as powerful and impressive as these highly intimate and obsessive imaginations promise to be. The object of desire almost necessarily must be massively disappointing—in contrast to your or my wet dreams about the actual fulfillment of this desire. The advertised wonders will never meet up with the actual mediated experience. This disappointment in usage though is a substantial and a recurring experience with new media gadgets or new media productions.

Sound design is—at least in this respect—no different from most of the other fields of media extravaganza. The disappointment with a certain sound design is almost programmed, the more the promise of the extraordinary is spread. Therefore, the propaganda, the advertisements and product presentations, the massively mediated narrations of the extraordinarily life-changing experience with a new sound design or media gadget, these are essential for the experience and the public acceptance of any media and any sound design:

> When Ford had designed a new fancy turn-signal sound ("pock-pock-pock") for its Ford Focus, many test subjects and automotive journalists initially rejected it because of their nostalgia for the venerable "clock-clack" sound originally linked to the relay. Only after the Ford Focus had been advertised as a first-rate modern car and a story had been linked to its futuristic turn-signal sound did the new sound come to be accepted. (Bijsterveld and Cleophas 2012: 116)

As in this case of automotive sound design it is the narration around and supporting this sonic product that helps generating acceptance and even a positive notion—in this case of being a futuristic, an agile, a desireable sound in contrast to the ancient and well-known turn-signal sound. The magic of sound again needed to be framed and narrated and staged. Sound designers and companies using sound design apparently need to work strongly against the substantial disappointment—either out of hyperbolic expectations or out of a too well domesticated and assimilated sound design that consumer citizens do not imagine it could ever be changed. Disappointment is hence a core element of every new mediated experience, and even more so in a media genre like sound design that still is regarded by the wider public as fascinating yet strange. The very impressive self-presentations of sound engineers and sound designers explicitely intend to work *against* precisely this lack of knowledge; though in the end it might be that some of these presentations only stabilize the image of sound design as being magical, almost unintelligible, and a domain of highly sophisticated, technical, acoustical, and compositional mastery of a superhuman quality. Such an image as a shamanist professional surely gives way to the wildest of fantasies concerning obscure everyday practices out of which sonic artifacts magically emerge. Undoubtedly this will also serve a certain business plan in the sound design field: being a magician or a superhuman inventor is by all means quite clever to attract followers, believers, and clients. This strategy of a trickster creates in the end though even bigger expectations and dreamworlds about what it could be that these magical designers would actually do—and what sound design can actually achieve? Is it really true—as the urban legend tries to make you believe—that the biggest automotive companies have hundreds, if not thousands of sound designers just to create the perfect slam of *one* car door? Is it true, yet another urban legend, that for a long time radio, TV, and movie commercials—and even ringtones!—have been engineered with arcane psychoactive techniques as developed by the US Army to access the subconscious of its sonic consumers? Is it true that the sound of crisps is also being thoroughly designed by sound engineers in huge and secret laboratories? Is it true that you become more productive—in the sense of ROI, the ominous *Return On Investments*—if the sounding environment you work in is engineered according to certain principles of acoustics and composition? This might not come to you as a surprise but if believed literally all of these urban legends can only create disappointment. The reality of the sonic experience in all of these cases as well as the reality of the sound design workflow behind it, the everyday *sonic labor*, is way less exciting, way less magic, way less arcane and obscure.

Actual sonic labor is more in accordance with recent research on the precarious situation of labor in the so-called *creative industries* than with the visions of magicians of the auditory (*cf.* e.g., Hesmondhalgh and Baker 2011). The everyday workflow of sonic labor is better characterized by tight deadlines, precarious freelance contracts with grotesquely delayed payments, scarce briefings that often are void of all aspects that might be relevant for designing a sound of a material hardware or software product, and sudden reversals of briefings and pitch-presentations that time and time again are requested to be the final product—or are requested to turn into something completely different, just because the CEO of the client's company likes a certain musical style or a musical instrument so much (most of these examples can also be found in the next part of this book: *Sonic Labor*). Notably, these characteristics are not specific to sound design alone, they apply to almost all branches of the freelance design labor market. With the big difference that the longer a design discipline is established, the more its genuine requirements and the respect demanded for the design labor have trickled down into everyday business: there is then at least a certain—if often tiny—chance that designers can work with clients who actually have a glimpse of what these designers *actually* do and what their specialized work actually does require from the side of the client. With sound design there is still a long way to go. Hence the disappointment is also one on the side of the clients and the producers—not only on the side of the customers.

The decentered media

The ubiquitous experience of mediated sound as guidance and escort has finally been established to a full extent in the early twenty-first century. The once rare and exceptional experience of a vast variety of sensory wonders, a multisensory shock, has turned into a continuous, and ever so often annoying if not harassing, company to everyday lives. What has not been lost though is the structure of disappointment, of anticlimax, of a frustration with the gap between a massively advertised, magic extravaganza and the effectively experienced, sloppy meagerness of media apparatuses or media productions. Indeed, both the voice assistant of the automotive navigation system that guided me earlier today—designed by one of the projects of our fieldwork—as well as the audio guide for an exhibition I visited months ago did not actually provide the promised *Elysium of Sound*. This substantial and very common disappointment apparently grew proportionally to the amount of technological apparatuses and transmitted productions disseminated all over the electrified, commodified, and networked areas on this planet. The longer consumer citizens like me and you live together with and in continuous exchange with media, the more it seems that the ambitions and desires connected to these entities have turned back against us. They seem to resemble, for some of their users and their consumers at least, more of an annoying dead weight— for them they seem to be less of a liberating, magical formula to lighten up their everyday lives. One reason for this frustration, this slight yet everyday experience of a hunch of suffering, might be found exactly in this ubiquity, this vast dissemination of media.

In the discovery period, in the inventors' and pioneers' period of media in the late nineteenth and early twentieth century, the situations of mediated sensory extravaganza were quite rare. They could occupy only selected if not actually very tiny volumes of our lives and bodies.

Moments of extravagant experiences, special events on Sundays or on holy days, on peculiar evenings or in the holidays, when one went to the church service or to experience fairground attractions, to celebrate extraordinary events, or to go to the theater, to the opera, to the cinema. Since then the sensory experience, the polysensory and the sonic experience with media has grown, has territorialized ever greater volumes, bodies and houses, worktime and leisure time, intimate and professional relationships have been continuously adapted to these new sensory materialities. These sonic products have been domesticated by us consumer citizens for quite some time now. These products themselves apparently did also make some rather impactful efforts to domesticate in return their incubators and providers.

This large and successively global and reciprocal process did play out, regarding a cultural theory of sound design, as a form of *sonic domestication*: today one expects certain notification sounds, ringtones, signal or alarm sounds, confirmation or rejection sounds as well as operational noises and helpful remarks from certain guidance avatars at almost *any* given moment. The inventors, designers, the engineers and law-makers, the selected users in the focus groups who tested these sound designs as well as some friends, partners, and colleagues who contributed on some detours to the finalizing of the individual sound designs, all these freelancers, employees, and volunteers did collaborate with the tools and machinery, the gadgets and state of software, with some design fads and game sounds to generate this sound continuum you and I live in now. The designed and sounding world has truly decentered from the state in which there was maybe a countable, rather single digit number of sound sources you would encounter on any given day. What once was a convenient, maybe extraordinary event in one's lives or days has become a ubiquitous commodity. As the lives of commodities in mediatized, capitalized and excessively economically structured societies go, very soon *after* their invention and market introduction they are not anymore considered as optionally or arbitrary lifestyle choices. They become required and almost essential attributes of a *consumer citizen:* a sensory and a *sonic persona* (Schulze 2018) who intends to partake in the most crucial social processes and who also wishes to be recognized as a serious, self-reflected, strong and able, a more or less sane, generally not criminal, and at all costs surely *not* a terrorist subject. But what indeed does turn a consumer citizen—as the fundamental figure in contemporary economies and nation states—into such a sonic persona, maybe capable of dealing with multiple forms of decentered media sound? Maybe there are even more traits or personae a listener, user, consumer could also perform? And would such a change of a sensory habitus just be a deliberate choice—or are there other transformations of lifestyle, perceptual techniques, lifeforms, or even forms of occupation that are involved in such a change?

Going back to the *anthropology of sound* (Schulze 2018) and its explorations on the various sensory forms of habitus, the perceptual habits and performativities of sensibility, I would like to introduce here in this book one often neglected and repressed role of sensory, reflexive, and social activity that can be considered a foundational role for various other personae: the role of the so-called *humanoid alien*—as introduced in *The Sonic Persona* (Schulze 2018: 8 and *passim*). *Humanoid aliens* belong—this is my definition—to the same species like you and me, more or less, and they show a wide ranging, material and experiential multitude of bodily shapes, skin colors, forms of habitus, clothing, body armor, and body enhancements of all sorts, like me and you. This multitude of appearances as well as its forms of existence, all the modes of sensory experience and perceptual idiosyncrasies are so wide-ranging that it really seems appropriate to call them humanoid aliens. This new term provides a variable, plastic, and malleable placeholder for an endless variation of diversities in order to prevent a hastily normalizing of body shapes, of idiosyncratic sensibilities, or performative preferences. A still very common but quite obviously reduced term like *The Human Being* cannot open up this wider variability: this older term refers by historical examples and established models of reference mainly to male, white, middle-aged, heterosexual, supposedly nondisabled and generally anthropoids, trained, skillful and healthy in all areas of life. A reference model of perfection, standardization, and flawless normality that almost never and nowhere ever exists or ever existed. Only if everybody is truly some different kind of alien (Schulze 2016b) one might at least somehow be able to understand how decentered media actually affect this or that humanoid. The diversity of decentered media can possibly be met at best by a diversity of decentered humanoids.

A *humanoid alien* might therefore show to a certain extent its qualities as a *sonic persona* and it might be regarded—in the sensory and sonic economies of the early twenty-first century—as a mere *sonic consumer* who needs to adapt to the expected behavior of the *citizen consumer*. All these four traits and personae are present in one's everyday life experience of functionally designed, ubiquitous sounds. These four personae represent aspects of our sensory existence and they also appear in mixtures and variants in our everyday lives. You and I, we might navigate—sonically and sensorially—in various moments, encounters, and affects between these (and many more) personae; we might blend, mingle, and shuffle the portions from each to try to act, to react, to perform, or to sense, to resonate, to activate in a certain situation of everyday life in a way we sense to be appropriate or necessary or the least wrong or irresponsible. In this respect, on the one side the user of a certain product with designed sound and on the other side the inventor and marketer of this sound design product are both, for instance, sonic consumers: they are compliant consumer citizens of this sensory economy. These four personae and how they explicate the wide varity of sensory experiences will be further explored, applied, and confronted with each other in the subsequent chapters of this book.

The decentered media and their decentered sounds that populate this solar satellite resulted over the course of the last handful of decades in ubiquitous media and ubiquitous sounds—as well as in ubiquitous forms of consumption. But is this notion of the ubiquitous (as explicated by Kassabian 2013, 2015) identical with a turn to the negative, a harassment, a sort of incessant punishment and dead weight? Actually, this ongoing process of decentering and ubiquifying resembles a process of dealienating and naturalizing anthropogenic inventions. New inventions lose their initial strangeness over time, their alien appearance: they become one of us, one with us, one like us. They live with us. We live with them. With new sounds and screens, membranes and pixels, pages and messages, environments and figures. The new media and new sounds become ubiquitously present like any commodity. To understand this it takes, on the one hand, a *non-anthropocentric anthropology* (Schulze 2016a, 2018): as the issues of an anthropology of sound can be better described and analyzed if the actions and intentions of humanoid aliens are not considered as the one and only center of developments, activities, and effects—this extends to the many cultural areas that constitute the vast field of a sonic anthropology; for instance, the field of a cultural theory of sound design. On the other hand, this also requires an understanding of *non-mediacentric media studies* (Morley 2009; Krajina, Moores, and Morley 2014): the issues of sound studies can apparently be better described and analyzed if the consequences and inherent agency of untamed, not yet domesticated media are *not* considered to be the one and only center of developments, activities, and effects. The process of decentering and ubiquifying extends, therefore, to research—to researchers, to their objects, areas or fields of research, to the forms of publications and presentations. Hence, this decentering and ubiquifying needs to be considered in research as a major ground and a major characteristic trait of this mingled and dynamized experience of sounds and their design.

3

Suffering

On contemporary sensory deprivation

The demand of connectivity

Humanoid aliens are connected. Not alone by means of their individual sensibilities, refined cultural knowledge, or sensory training, but of course as *consumer citizens* by means of the various and ubiquitous electronic networks connecting all areas of everyday life. The online administration service of the municipality I live in has its sonic signals as well as the public transport system that I use for commuting and the streaming or radio service I listen to while commuting: and all three are designed by freelancers who are employed by communication agencies or directly by larger corporations. *Sound design facilitates connectivity.* This materially achieved connectivity has been as selective in its earlier stages as the aforementioned forms of multimedia extravaganza were in their times. This too has changed, it evolved, multiplied, and has tended to also become ubiquitous, ceaseless, and seamless. The mere fact of connectivity has a tendency to vanish and to invisibilize itself like any fundamental resource for living and existence, like any of the *common goods* or the *allmende* (Kauls, Grunberg, and Stern 1999; Bollier 2014): the air you breathe, the water you drink, the sounds you hear. At the same time exactly this ubiquitous and continuous character of being physically connected seems to lead to an even stronger materialization of feeling connected, a sense of almost being chained to one's accounts, one's communication threads, the companies and platforms with which one engages: a ubiquitous and continuous prison in the form of tracking one's metadata, of storing one's full activity protocol (Boever and Neidich 2013; Neidich 2014, 2017a, 2017b). Is one's existence now reduced to just a lifelong sentence in the global jailhouse of media and consumption? Is it anyhow possible to really live as a *platform agnostic*? The whole apparatus of material objects, of technologies of image, movement, and sound recording, processing, transmitting, and reproduction, can be understood as a huge imaginary that has materialized so powerfully in our cultures and is hugely present everywhere. In rare or not so rare moments of our lives we might then indulge in moments of our own individual boredom, our deep intellectual doubts, and we might experience what I am tempted to call: *the user's headache.* In such periods of ache and suffering, of momentary rejection of this apparatus' demands—which seemingly not so few people working intensively with audio technology experience—we might even be tempted to follow the ironically relativist, the joyfully destructive, and generatively conspirational thinking of a subculture book like the *Principia Discordia* (1965). We might like to call this armoring apparatus imprisoning as a *Black Iron Prison* that has been built around us. *The Black Box* of technology, so we like to believe, has grown into everyone's *Black Iron Prison*. There is finally no way out. We ourselves are the most stupid servomechanisms of this sealed technological capsule we built. Our very own sensologies hold us in bondage (Schulze 2016a).

One form this black iron prison of ubiquitous connectivity takes is obviously the form of functional sounds and guidance that are confirming or denying access to selected areas and that are guiding your way through a given system of access points according to the guiding principles of the software in question. One day in a life can take the shape of constantly being pinched and perforated by the various sources of functional sounds. Any consumer citizen can live under the impression that their individual life is only worth as much as their consumption activity is acknowledged and evaluated at by the relevant authorities—state authorities and corporate authorities (Aranda, Wood, and Vidokle, 2015). This impression is grounded in the fact of ubiquitous data collection, in the extraction of valuable interpretations from metatags by tracking and in the networked databases that allow for a comparative evaluation. Whereas these activities in the transformation of material activities into rather less material datasets are a given fact—yet presumably far less systematic and malevolent and consistent than often assumed—the interpretation as constituting a continuous jailhouse is not. At this point of this study it is though safe to state: this impression is a dominant interpretation of contemporary everyday usage of connected media.

Regarding its connection to sound design and to a growing amount of functional sounds of admission and rejection, of guidance, navigation, and expertise, this interpretation receives more and more arguments and examples of its impact, its everyday annoyance, and continuous paternalism in all areas of life. This proliferation of functional sounds into any given moment and instance of one's life contributes to the aforementioned impression of a continuous auditory surveillance in the *Panacousticon* (Schoon 2012). The main characteristic of this ongoing auditory signaling of admitting or rejecting a certain consumer citizen is its pressing, almost juridically effective demand for explication of actions, intentions, trajectories. The recent processes of technologically implemented gating of mostly urban areas allow hardly for an admission that is *unsecure ...*, *Just trying!*, *Maybe?*, seeking camouflage, or hiding in timid shyness. Any humanoid alien experiences itself as encapsulated in such habits or as seeking such forms of habitus in certain moments of their life. Yet, such a behavior of insecurity and uncertainty seems almost impossible to execute if one is immediately tracked, recorded, and scrutinzied. Timid, implicit, and cryptic activities are excluded. They are—regarding the request for explicitness by these gating procedures—almost logically inexistent. Explication of one's actions toward the sources of functional sounds is crucial for the existence of consumer citizens in the twenty-first century. But how can one live with a constant ratification of one's actions? Does this not establish another, quite radical change in the relation of me to my individual actions? Am I not forced now to regard my very own actions as being outside of myself? As if they were material performances of activity every time? Does this incessant ratification of action not lead to an even more and even much more massive alienation from one's own actions? Yet, maybe this alienation is the best thing that could happen to the consumer citizen? Isn't it a memento to you and to me that we are not regarded foremost as individual aliens and beings but as accounted consumers whose consumption only grants us essentially our citizenship?

This throughly explicit character of admission and inspection procedures, framed by functional sounds, is an example of the concept of *sensology*, introduced in the previous chapter (in the section "The sonic consumer"). This process is part of a general sensology as elements of a disputable ideology are implemented into everyday life not by official, legislative decision and preceded by public discourse as it would be the case according to theories postulating a forming of the political will; in these new cases though elements of ideology are concealed, they are sneaking subtly into everyday life by lower, more administrative decisions, considered not relevant for legislation as they are mainly obtained by individual, mostly corporate actors. They remain concealed as of lower relevance, as only administrative, and only temporary until they are fully installed and not seldomly executed as if they were implemented after a higher federal order if not executing international law. This proliferation of everyday life is a symptom of *sensological transformations*: what comes at first as a temporary fad, just a momentary impulse, soon to be forgotten, suddenly is implemented and maintained as a now permanent structure, requiring obedience by all consumer citizens affected by it, with no sign that this sensological tool might ever vanish again. In the end the individual life on every single day has changed, affecting every single trivial action of using public transport, of entering a citizen center, of asking for access to one's own bank account, to one's own healthcare records, to one's own credit score records, to one's own tracked metadata of various mobile devices. The *security theater* (Schneier 2003) in particular and the *securitization* (Buzan, Wæver, and Wilde 1998; Balzacq 2011) discourse in general has become ubiquitous, oozing out into every last crack of your and my individual life. The great gift of ubiquitous connectivity carried, in the long run, the pathogenic poison of permanent explication and self-documentation.

Rigor mortis

Sound design in the early twenty-first century is to a large extent the design of functional sounds, sticking to specific appliances, gadgets, and products, and most of the time part of the aforementioned admission and inspection procedures. This holds true for a long series of projects we encountered and partook in our fieldwork, beginning with the sounds for an urban public transport system, the sounds for a newly developed mobile phone outside of the duopole of Android and iOS, the sound branding of a new development quarter in a metropole city, to the sound design for a municipality's online administration system. The structuring process of explicating individual actions in a web of ubiquitous and ceaseless connectivity transforms in all of these cases the interactions that humanoid aliens need to perform. Actually, they are no longer regarded as humanoid aliens with their idiosyncratic preferences, obsessions, quirkiness, or shyness; obviously, you or I are also not regarded as a *sonic persona* with a *sensory corpus*—that would also imply rather complex forms of existence in a humanoid alien. But the major processes of admission and inspection, also the act of inspecting and admitting the activity of gadgets or appliances, all of this requires another habitus, another persona: the persona of a *consumer citizen*, recurrently mentioned already on the previous pages of this book. As probably has become clear by now, this notion of the consumer citizen is far away from an affirmative or even positive, empowering concept. The consumer citizen is—in comparison to concepts of a person as a humanoid alien, a sonic persona, or a sensory corpus—more of a rather functional if not sad reduction in pragmatics. This reduction is moreover one of theory: it is in the responsibility of the author of this chapter. Yet, this precise reduction is one that is employed at the same time in the very empirical field as observed by the author. You might hear now a subtle, gray, and clicking login sound.

In the context of an anthropology of sound this habitual reduction is not a neglectable accidence: it is a substantial transformation—if not disruption. The humanoid alien as a malleable and plastic lifeform investigated in *The Sonic Persona* seems to play a minor role in contemporary processes of admission if it is not radically erased from the admission process. If you start dancing or yodeling in the process of a security control you will surely be expelled if not arrested, maybe your documents of identification are taken from you, maybe you will not even be granted a proper court case? Such actions are considered as idiosyncratic—yet more in a diagnostic than respectful way: diagnosis of a certain pathology or a criminal action. A sonic persona or a sensory corpus might behave in such unforeseeable ways though. But as soon as some subject follows such a behavior, as soon as you would follow *sonic* zones of pressure and desire, rejection or attraction, as soon as I would rely solely on my *sensory* preferences and needs, hurts and trajectories, we both would be expelled, we would evade the categories. We would not exist. The hegemonic tools of admission might at this point of history not be able to deal with some subject presenting individual, idiosyncratic obsessions and fears, hopes and desires, behavioral ticks or semantic swirls. These mechanisms of inspection and admission seem to be lost—or at least they neglect investigating any further, they neglect any admission at all. A sensory corpus is not even existing as a category in these investigations for admissions and the same is true for a sonic persona or a humanoid alien. They are no subjects: they are *nonsubjects*. A total sensory restraint and deprivation of the senses takes place.

The only subject recognized is factually the *consumer citizen*. You hear the confirming sound that your credit card has been accepted via an electronic connection to the bank issuing this card. This persona is granted access, a form of access though that is highly formatted, restricted and—as already mentioned—massively reduced in comparison to the vast range of activities that are anthropologically possible or probable for a humanoid alien like your or me. But what is the main difference between these two roles—could it be that it is just a certain intention or a more superficial attribution of activities and qualities that leads to this distinction? Maybe it represents only a personal if not solely idiosyncratic preference? I hear drones, deep drones of the resonance of a subway train, rolling by beneath my feet. Could this really guide the way to substantial, anthropological distinctions? Surely not. Yet, the distinction is anything but merely idiosyncratic or superficial. This distinction is substantial, it has its roots in the anthropological concept underlying certain activities—for example, processes of admission and inspection. It relies not on attributed tags or qualities, but on inherent characteristics of action and performativity. To unfold this, it is necessary to refer to the fundamental qualities of these different personae. Whereas the persona of a *humanoid alien* is characterized by its broadness, its plasticity, its viscosity, and its almost endless options of variations and alterations, this is not true for a *sonic persona* and its *sensory corpus*; both of which are focused in their activities around sonic expressions, radiations, and sensibilities, respectively focused around sensorial, material, and physiological effects and actions. Furthermore, the *sonic consumer* is rather limited: as this persona is focused around the sonic, yet not as a means of individual and personal expression, existence, or performativity but as an executable form of consumption which encompasses all relevant forms of activities of this persona. Finally, the *consumer citizen* now in this range of personae becomes recognizable as even more reduced: reduced to forms of consumption—but even excluding or not especially focusing on the sonic. Consumption is the raison d'être of this persona. Another login sound. Without forms of consumption its existence would neither be thinkable nor viable. An existence that hence does foremost require actions of consumption does not require actions of sensory expression or sensibility. These actions are contrary, even possibly considered distracting, as they are truly not solely restricted to activities of consumption.

Now, in the process of consumption any other activity occurring is also reduced to consumption only. This is also true for the process of admission and inspection. In consequence, the most appropriate and requested behavior is one that shows the least sensibilities or affects, the least kind of expressions or idiosyncrasies. Sensory deprivation and restraint are basically requested to get any admission. It seems to be absolutely necessary that in the process of admission and inspection the lifeform in question is not showing any tensions, insecurities, ambivalences, or extravagant seizures: not in speaking or moving, not in walking or standing, not in strolling or gesturing. All of these kinds of behavior would be considered as potentially harming as they draw the subject in question out of focus—and any set of criteria can then hardly be applied or investigated. It is demanded that a subject that wants to be admitted needs to behave like they are in *rigor mortis: playing dead.*

> A body is therefore a tension. And the Greek origin of the word is tonos, "tone." A body is a tone. I don't say anything here that an anatomist couldn't agree with: a body is a tonus. When the body is no longer alive, has no more tonus, it either passes into rigor mortis (cadaverous rigidity), or into the inconsistency of rotting. Being a body is being a certain tone, a certain tension. I'd also even say that a tension is also a tending. (Nancy 2008: 134)

It is this rigor mortis body posture, without individual desire or idiosyncratic tension, that needs to be assumed as soon as one is entering a process of admission and inspection. Any idiosyncratic or individual behavior needs to be considered in such situations as harmful, as irritating, as despicable. It is required to behave or at least to simulate the existence of "a kind of statue or automaton with twenty layers of armour, a veritable Carpathian castle" (Serres 2008: 145). Only if the state of rigor mortis, of being a statue, is achieved, can the admission process start. Then you will be investigated. If you dance to the scanners or blinking lights, to some alarm sounds or people chatting, if you engage in chatting yourself then you are out of the game. You are excluded and not admitted. Your—luckily temporary and habitually performed—state of rigor mortis is required for you to get access. This is not a metaphor, this is not some ironic twist or interpretation. This is an actual description of the behavior of a consumer citizen in the face of surveillance and control, in the process of being scrutinized by an authority. Sensory deprivation and restraint is the consumer citizen's constant guard. One is required physically to act as if dead: rigor mortis is the only granted savoir vivre in times of pervasive and ubiquitous surveillance.

Ideologies of efficiency

The persona of a consumer citizen is—obviously—a polemic sketch. None of you or I would probably agree to being such a kind of person all of the time and in every instance of our lives. Yet, there are enough moments—be it at the workplace, whilst shopping, while gleefully or thoroughly annoyed *Wasting Time on the Internet* (Goldsmith 2016), or when passing the now repeatedly mentioned processes of admission and inspection at airports, at corporate or state official edifices—in which you or I can sense: *Now I am nothing more than the proverbial cattle being lead from one point to another. I am no one. No one cares for or is even interested in my individual needs or desires, idiosyncrasies or longings. I am irrelevant.* The consumer citizen is the name for this irrelevant persona, this troublesome entity, this soft machine to be processed. A processing that is confirmed, escorted, embodied, and memorized by the occurance of functional sounds.

The eminently nonrelevant personae, be it a humanoid alien, a sensory corpus, or a sonic persona, do actually die or at least hide and cover if treated in such a way. The sonic consumer though, being foremost a consumer citizen, indulges in these activities as they grant them access to sound products and to processes of consumption. Me, also being such a sonic consumer, I follow these guidelines, I long for their promises, I obviously will be obedient enough to access the selected product purchase options and to follow the instructions given to me for correct and planned consumption. I consume these sensory products of sound design and sound production the same way you do: when we buy an appliance or a smartphone app, if we enter a building or an intensely digitized nation state. As a consumer citizen one truly accepts all these requests and demands of the enveloping dispositive of capitalization, purchase and consumption.

One might ask: why is all of this critique of society, of ideologies, of work conditions or of the security theatre globally running wild since the announcement of the GWOT (Global War Against Terrorism) in 2001 of any relevance for a cultural theory of sound design? Yet, in asking this question the asker implies that insight in one specific area of cultural practice is only to be gained whilst completely ignoring how it is connected to or determined by realms supposedly outside of it. But what area of cultural practice is truly unrelated to economic, to juridical, to political, to administrative areas of culture? The contrary is true. It is a necessary prerequisite and also a methodological axiom for cultural research: in order to understand one tiny, incredibly specialized area in the life of humanoid aliens one must at the same time indulge in the specific practices, the self-perceptions, and all the characteristic activities of this area—as well one must also trace the threads and layers of culture rooted in areas of cultural practices outside this rather narrow field. Only then one can hope to achieve not only an immersive and possibly surprising account of the individual practices, narrations, imaginations, and personae but also how these are related to ambient areas of cultural practice. This main insight of a *historical anthropology* (Wulf 1997, 2013) guides this research here in our strife for comprehensive interpretations combined with detailed, narrowly situated analysis. This holds especially true for professional practices and their highly refined practices, carrying with them an immense historical, contextual, and pragmatic load.

Hence, in order to research the cultural practices of sound design one must necessarily understand the fundamental traits of its hegemonic, encompassing framework of consumption and surveillance in the early twenty-first century. Otherwise this research would confine itself to a form of willful myopia: a myopia of one limited area that is the exact opposite of focusing. If one researches the impact and the limitations, the generative potential and the depressing restraints of sound production in the field of design, then the economic dispositives are as important as the political and administrative directives bullying it: "The flaneur died with the birth of the department store" (Goldsmith 2016: 66). There are such *ideologies of efficiency,* as one might tend to call them. Such ideologies are manifest in these dispositives and directives, they materialized in contemporary culture mainly in the form of tendencies toward an overload of surveillance measures and of an almost incredible economization of the singlemost area of everyday life, in historical comparison. They intend to accelerate and to increase the degree of efficiency in possibly all areas of life—and in the end they seem to increase mainly the amount of surveillance measures and of economized areas. The territory of the lives of humanoid aliens like you or me is occupied and colonized, step by step, by forces intending to achieve a balanced and monetized fiscal economy and a seamless sequence of tracking and stalking, of permanent and comprehensive surveillance in every single point in time and space, of every single actor in a given group of consumer citizens—under the claim of their own benefits, understandably.

One of the historical roots of this complex amalgamation of ideological assumptions can be found in the science history and the cultural history of underlying notions of effectivity, of logic, and of concise and direct communication. One of the theoretical concepts underlying such assumptions is surely to be found in information theory as it was developed in the Second World War. In the framework of an anthropology of sound (Schulze 2018) it becomes quite clear that these theories of communication, of signal transduction, and of information need to be regarded as historical constructs, rooted in the history of warfare and political struggles of the twentieth century. Under the influence of a war economy and its efforts to merely survive and to use all the resources at its best, this definition, as well as the whole theoretical apparatus of information theory developed and flourished. It carries, though, the epistemological birthmarks of a historical period in which shortage was the major experience for many citizens in Europe—and it was not so remote an experience for any military personnel on duty in the United States. In all of this, an imaginary of military engineering seems to prevail. This might well be appropriate for a lifeform under a military dictatorship or a state in permanent, infinite war. But this field of military activities cannot be taken seriously as a general example of how humanoid aliens do lead their lives under all circumstances. It can serve as an example for military organization itself, for life and culture in wartimes—but it would be arrogant, preposterous, and somewhat ignorant to assume this social field could serve as a model for social interactions in almost every other field of humanoid lives (even if the *militarization* and *securitization* of everyday life in the late twentieth century seems to insinuate such a trajectory: e.g., Gillis 1998; Hogan 1998; Buzan, Wæver, and Wilde 1998; McEnaney 2000; Roland 2001; Balzacq 2011). And yet, exactly this insane transfer happened by way of the engineering imaginary becoming the fundamental condition of life in networked societies. Militarization is an underlying blueprint for activities in the areas of economy, of politics, of communication, even in design and the arts, in music, and in performance arts. Militarized reduction as a model of connecting has become hegemonic. In the early twenty-first century it has become increasingly difficult (in some areas close to impossible) to describe, to understand, and to analyze the relations and activities of humanoid aliens *without* any direct or indirect reference to military organizations.

Hence, the militarization of consumer culture and everyday lives results foremost in a pervasive wish to achieve the most effective and the most efficient way of living, of consuming, of communicating, and of production. This reduction is indeed an effect of technological dispositives inherently representing military organizations and relations. Thus, the whole cultural and social environment is gravitating toward military organizational structures and working processes, toward military forms of continuous and explicit documentation and protocolling as well as toward military relations of consumer citizens in their individual activities of communicating, checking in, sending, receiving, and confirming. Society is mimicking the barracks.

Imploding idealizations

Where do these situated investigations and anthropological reflections regarding the degree of suffering in producing or consuming sound design after all lead to? Don't they actually lead to nothing more than just reproducing contemporary, rather whimsical rants and fears concerning the new apparatuses, the new cultural practices of tracking, of being ceaselessly online and present, and of in-depth data research that now populate this globe? Even if we encountered again and again in our fieldwork this well concealed suffering as a continuous bassline running through all meeting sequences, milestone revisions, version updates, contract renegotiations, and quarrels about licensing agreements in the process of developing, finalizing, and implementing a sound design, are you or am I actually experiencing this sensory deprivation and restraint I was claiming as dominant in contemporary societies, are we experiencing this every single day? Don't we feel liberated and elevated, excited and aroused by sounds all the time? How could all this really relate to the common practices of inventing, implementing, and digesting sound design?

Contemporary disdain for specific cultural practices is first of all not irrelevant for the status of these cultural practices themselves. On the contrary, they also get shaped by those assumptions, fears, hopes, and obsessions as well as by the urge to counter or even to overcome contemporary attitudes toward one field of practice, shaping its cultural status and cultural impact. Therefore, the function one imagines and ascribes to sound design is not irrelevant to its actual, material practices—even if these imaginations might have almost no relation to the pragmatic daily sonic labor executed.

In the case of sound design the common imagination of anyone making music, indulging in music, or creating sounds or sound design still contains to a large part in the early twenty-first century assumptions about musicians, entertainers, and theatre ensembles that already were present in Western cultures at least since medieval times and in part already dominant in ancient Greek and Roman philosophy. Assumptions still run wild about a loose and irresponsible kind of people, too lazy for actual work in comparison to more than your common bourgeois patron and too irresponsible and ignorant, too playful and dishonest in comparison to all the proper craftsmen, tradespeople, and businessmen—proving Jacques Attali's early essayistic diagnosis right:

> The musician, like music, is ambiguous. He plays a double game. He is simultaneously *musicus* and *cantor*, reproducer and prophet. If an outcast, he sees society in a political light. If accepted, he is its historian, the reflection of its deepest values. He speaks of society and he speaks against it. This duality was already present before capital arrived to impose its own roles and prohibitions. The division between *musician* and *nonmusician*—which separates the group from the speech of the sorcerer—undoubtedly represents one of the very first divisions of labor. (Attali 1985: 12; first pair of emphasis in the original, second pair my emphasis)

Exactly this division—between "musician and nonmusician"—establishes one of the major idealizations in the history of music and of sound; similarly as the assumed laziness, the irresponsible corruption and slouchiness, the malevolent and greedy imposture of certain patrons and tradespeople is just another cliché as well. So, shall one conclude that all humanoid aliens in business relations turn to their ugliest and worst own avatar? And should they just play music and become composers to bring out the best in them again? Is this the vulgar anti-salesman- and pro-artist-ideology one should follow? An ideology that is at the core of a bourgeois division of labor to keep areas of society and of activity neatly separated and nicely controllable.

Instead of reproducing such negative assumptions and neatly separated clichés of a profession—that always include and hide the most positive and euphoric clichés and assumptions about this profession as well—instead of perpetuating such unfounded and self-contained rants and ramblings, it seems far more useful and insightful to encounter *actual* practitioners and to escort them whilst performing their actual practices. This still will include a large amount of idealization and self-stylization—but then there is at least an option for reflecting upon this bias: the resistance in actual encounters.

The clash between idealized, prefabricated imaginations of a field of practice and the actual, often monotonous, in parts repetitive as well as often annoyingly insecure and precarious conditions of labor is a recurring conflict when researching the sonic labor of sound designers. This clash was the nucleus and the main impetus of the research behind writing this book: Why are such weird and largely unfounded assumptions about sound design still rampant? And why is there no comprehensive cultural theory that locates actual sound design-practices in the wider realm of cultural practices, of design labor, and of sound production? Why are there recurrently newspaper articles highlighting the magic of sound design as well as manuals on handling the relevant soft- and hardware, but barely any comprehensive reflections on the aesthetics, the cultural value, and the political or social impact of this sonic labor? As soon as one dives into the sinuous practices of this professional field one actually does encounter a wide range of approaches to sound design with backgrounds in instrumental practice or in visual design, in music production or in programming, and in architecture or in advertising that we will lay out in the second part of this book, *Sonic Labor*, in visual form. One also encounters the ongoing hope and desire, the strife and the urge to craft a sound design that truly opens up a gate into something exciting: an excessive jump into yet unfamiliar, unknown, unthinkable sonic territories. A deep longing for this sonic joy, also catered by being bored with existing specimen of sound design. Where is the next sound? Where can I find some better design?

But the pressing questions that push designers forward do not stop here. They keep on bugging: When one suffers with existing design of sounds, one tries at least to provide some better examples oneself, right? How could this be possible? What would one need to overcome to achieve this goal? In what ways could the present conditions of working on sound designs be altered in order to achieve another kind, maybe a better kind, at least a different kind, of more intriguing, of more appropriately shaped sound design? But how could this actually be done? Are the conditions of sonic labor not invariably fixed and have the individual sound designers not too little room to maneuver or to transform these labor conditions? What can a sound designer do? Are there any promising, realistic utopias around for how sound—and, in general, design—could be crafted in a more meaningful, in a more intrinsically rewarding way?

All professionals know very well this everyday tension between high aspiring dreams and urges and the all too often depressing reality of labor. In the end this might lead to a neglect, to a defeat, and to a concession that might give up all actual and original goals in order to subject itself to the ruling procedures of—in the case of sound design—the communication business, the creative class, and its humiliatingly low pay, recognition, and perspective regarding any greater success. Is this a kind of deathwish? Isn't it actually suicidal to subordinate oneself to such preexisting structures, demands, routines, and dispositives? But is there actually an alternative way? Isn't it true that ruling dispositives are always stronger than any individual, personal wish or urge for change, for transformation, for a better, a different dispositive? How can we change the conditions of sonic labor in the twenty-first century, and in what direction, anyway?

4

Joy

Bored by tricks and the lust for excess

Sonic excess (spinning away)

I am lost. The sounds are spiraling around me, they take me with them. I feel held and at home, yet lost and I am losing myself and this situation even more every second. Am I spinning away? "One by one, all the stars appear, the great winds of the planet spiral in" (Cale and Eno 1990: track 6). This is not just a sonic experience anymore—it is a sensory, a corporeal experience. I feel my listening body, my *sonic corpus* (Schulze 2018) activated, present, right here, right now. A form of excess is taking place that takes me out of my place, into a different area, a different *sonic fiction* (Eshun 1998), an inspiring, an invigorating, an energetic new *sonic continuum*. "Spinning away, like the night sky at Arles. In the million insect storm, the constellations form" (Cale and Eno 1990: track 6). I enjoy this experience, more deeply than I could confess right now. I feel as if I would be more present, more alive, more recognized and more vital than at other times. These sounds transform me—they excavate me. I am alive and this experience evades the usual effect one might expect today from sound design. Yet, this is exactly what sound design still does promise: it intends to provide and promote an excess into yet unfamiliar, unknown, unthinkable yet very agreeable, exciting, and excessive sonic territories. But why, when this goal is so often stated as the main task for sound design, when to achieve this kind of sonic magic, this audio surprise, seems to be sound design's maximum goal has it so obviously been lost? Why does sound design provide less and less of sonic excess, audio surprise, of sounding magic? And why does sound design seem more and more like an accomplice of pervasively expanding control and consumer societies?

"On a hill, under a raven sky I have no idea exactly what I've drawn" (Cale and Eno 1990: track 6). Surely, you are wondering why these words by John Cale and Brian Eno appear on these pages. "Some kind of change, some kind of spinning away. With every single line moving further out in time" (Cale and Eno 1990). They spin you away, I am spinning away. I indulge in these sounds, I do not really, not actually, realize where they might guide me to—but I love, I crave, I long for listening again and again to this one song, its sounds, its sound design, its rather idiosyncratic and yet extremely well-made sonic figurations. This craving is characteristic for an intriguing sonic experience. The *sonic dominance* plays out and your individual sonic corpus is activated, is drawn into following this music, these sonic arcs and ruptures. You and I, we then just follow them. We do not question too much any agencies or sources, styles or distinctions. Though of course we would love to do so, and very often we actually do. But now, we just follow our corporeal listening desires, our joy to listen, our indulgement to let our bodies follow these resonances and pressure waves, these basses and breaks, evolving lines and transforming harmonies, changing styles and traditions, instrumentation and sound designs. We love this. We enjoy this. We enjoy being sonic consumers in the best sense. We let our bodies become the arenas in which sound is taking place: "My soma is the sole arena of these sound events. ... The sonic arena is my body" (Schulze 2018: 212, 224).

This excess is a sonic form of excess. In this special case of excess, it is actually—referring to traditional signal transduction theory—not even necessary for scarce signals to be transmitted. This joy, this excess is a *performative excess in the realm of the sonic*. This is what the magic trick actually promises. This is the promise actually realized. Yet, how often does this take place? It seems almost ridiculous if one would truly intend to list all the moments when sound design indeed *did* provide this form of sonic excess. Maybe in a massive blockbuster movie from your childhood? Maybe a significant signal sound in the years of adolescence, waiting for a message, a call, a note, a contact request from your desired and adored significant-other-to-be significant other of your future life? Maybe in countries or countrysides you visited during a holiday or a brief getaway? Maybe the sound design of a truly and shockingly new gadget that invaded your household and your workplace by storm—and you actually somewhat enjoyed this? This list surely is insufficient but maybe it highlights that the joy in functional sound design is a rare though common event. And yet it is still and ever again the explicitly desired goal of sound design. So, must there not be a way to achieve this in the framework of contemporary cultural practices of sound design? Couldn't it be possible that the joyful spores and the desirable radiation of sound design just lie buried under the many annoying and harassing details of contemporary work cultures with an inclination toward ubiquitous surveillance and segregation in the so-called *creative industries*, held hostage by (or in) the *creative class*?

"And now as the pale moon rides (in the stars). Her form in my pale blue lines (in the stars)" (Cale and Eno 1990: track 6). Would I really seek this? Would I really be able to bear this? Would I really be capable of enduring a sound design around me, in my apartment, my automotive individual transport device, my outside clothes, raincoats, or umbrellas, my kitchen appliances or bathroom devices, if all of these objects would emit sounds that did excite me incessantly and guide me into forms of sonic excess? With no doubt this would be the wet fantasy of any propagandist of sound design on a globally in realtime transmitted major presentation stadium. Oh, well. No one probably could truly bear such an expectation of total presence, insistence, radical intensity and excess. *Total immersion* is truly never actually the goal (Schrimshaw 2017). It would simply break up the framework of everyday life—or is this the actual goal?

"And there, as the world rolls round (in the stars). I draw, but the lines move round (in the stars") (Cale and Eno 1990: track 6). Apparently, it is not all about joy and certain sentiments of feeling elevated or satisfied; it could be more about an integrated and substantial form of joy that would be radiating in certain situations. It could also be more about a situation at the workplace, at home, or in urban commuting that is not dominated by sound signals of alarm, rejection, or affirmation, of penalty or reward, of control or advertisement, but by sounds with a wider variety of qualities, maybe even with more respect, more interest, more inclination toward the humanoid aliens being their listeners. Intentionally, right here I do not use the words *consumer citizen* or *sonic consumer*; it just might be that exactly these two contemporary personae are a source for the missing joy, the missing satisfaction. The deep connection between *capitalism and depression* is irrefutable in the early twenty-first century (Fisher 2014). Hence, a critique of sound culture in commodified and consumerist lifestyles these days needs to be founded on a more substantial critique of exactly these lifestyles of consumerism and commodification. It would not be sufficient, maybe it would even be rather harmful, if a focus solely on sound production would just result in even more brilliant, more exciting, more intriguing, addictive and excessive sounds. In the end this would foremost contribute to a stabilization and an affirmation of major characteristics in contemporary capitalist consumer cultures—and in sonic labor especially. "There, as the great wheels blaze (in the stars). I draw, but my drawing fades (in the stars)" (Cale and Eno 1990: track 6).

A critique of sonic consumer culture definitely needs to play a major part in this book's discussion of a cultural theory of sound design. This critique though needs to be founded on the economic, the political, the cultural, and—bringing all this together—the institutional, educational, and juridical frameworks and traditions providing the habits and personae, the practices on media stages, and the processes of commodification as well as the dominance of capitalization and consumption that are characteristic for your and for my everyday life these days. It would never be sufficient only to tweak and fiddle at the most outer fringe of this immensely large supermetamegainfrastructure that has been implemented into everyone's life in recent decades and centuries and that one might legitimately call *Capitalism*. We—you and I—are living in a supermetamegainfrastructure that is more of a *Capitalocene* (Moore 2014a, 2014b; Szepanski 2014a, 2014b; Schulze 2018) than of an *Anthropocene*: not the needs, desires, demands, obsessions, and idiosyncrasies of humanoid aliens as a joyful multitude are the main reasons for earth's transforming activities in recent decades and centuries to come, but, more precisely, the needs, desires, demands, obsessions, and idiosyncrasies of capital to territorialize, to colonize, to multiply, and to occupy at least every single nanometer and nanosecond on this very planet (we might still have mercy for all the other planets and lifeforms in the vast rest of the space-time-continuum known to us; but not for long). It could well be that the urge for excess—and be it *sonic excess*—is more in tune with the anthropological concepts provided by this supermetamegainfrastructure than one might think. "And now, as the old sun dies (in the stars). I draw, and the four winds sigh (in the stars)" (Cale and Eno 1990: track 6).

Drifting, sensorially

Once in excess one might not at all realize what precise types of sound events or sound signals are actually present in a given moment. Driving though an unknown city or country, following traffic signs and sounds, chatting with your beloved, and listening to some favorite music, you might remember this perfectly shareable moment sonically. In moments of excess the sensory starting point for this drive into excess is surely lost—and this exactly *is* one of the joyful experiences in any moment of sensory excess. Yet this is not the only thinkable form of joy one could experience when encountering products of sound design. The joy in listening to such sounds is also an effect of a situated, a performative, a highly corporeal, a sensory and highly sensible event. Sound design affects its listeners in moments of activity, of layered intentions, and complex agencies, interwoven with sometimes contradicting or distracting side- or counteractivities lingering below or running on underneath. You make phone calls, receive video conferencing requests, you reply to instant messengers or chat requests, all signaled by certain sounds. It is almost never a sole and focused, an exclusive activity—only maybe in the first minutes and hours when learning the capabilities and the functioning of a new gadget, apparatus, or appliance. How does this water heater actually work? Did I set-up this phone correctly? Is my navigation system equipped with all necessary maps? It is always a mingled, dynamized, a multiple and complexly layered, very often even a lo-fi listening experience. An experience in which one simply has no other choice than to immerse oneself. If one would reject immersion in such situations of everyday life, one only ends up being considered a socially awkward and a weird outkast incapable of managing simplest everyday tasks. Immersion might be intensely desired, crafted, also discussed and doubted in artistic areas—yet it is the only really possible state of activity if living everyday lives outside of artistic practices. Immersion is dominant anyhow, be it in excess or otherwise, when living with functional sounds: it is just a combination of everyday life and the genuine sonic dominance. A permanent sonic excess and its specific forms of immersion hence cannot be the final answer to the missing joy in everyday sound design. There need to be other options, other ways of designing and implementing sound.

In situations of everyday life it is often a continuous but more often rather cut up experience of pervasive immersion that might first of all allow an experience of joy to take place. This experience depends not so much on the individual constituents of such a situation but on the activities, the relations, the expectations, and remanence inhabiting this situation. I come from a conversation, open a certain document, probably with a sound, I try to transcribe—with other sounds resulting from my activity—the outcome of our talks and discussions and enter it into this file. More sounds ensue. Almost the same sound sources might be perceived as harassing, as overloading, as inspiring or trance inducing, as danceable or uplifting, or as subversive or annoying. The generation of these affects regarding some sound events cannot solely be found in the sonic materiality itself—though this will provide a first crucial step. This insight into the material and subsequent effects of sound—as proposed by sound studies—the sequence of sonic experiences in everyday life are crucial. Sound design cannot be analyzed nor can it be evaluated and criticized if it is taken out of context and analyzed mainly in reference to historical predecessors, to musical or acoustical knowledge, or to technical requirements for producing and reproducing *only*. This approach to sound design in particular—and to sound events in general—I like to call the approach of *sonic transcendence* or *sonic metaphysics*. This approach stands in contrast to recent theoretical approaches of *sonic materialism*. Whereas sonic materialism (Henriques 2011, Thompson and Biddle 2013, Schrimshaw 2017) integrates as many material, physical, experiential, also historical, semiotic, psychological, and phantasmagorical threads in experiencing sound this other, more common and pre-sound studies approach of *sonic transcendentalism* does exactly the opposite. An interpretation according to sonic transcendentalism does not focus on sound precisely in its experiential environment but it loves to jump immediately, with no further ado, to some transcendent knowledge outside the actual situation of listening and sounding: be it of aesthetics, of acoustical physics, even of religious and historical traditions. However, these forms of knowledge, of major importance in their genuine area of discovery and application, then get to be stretched beyond their genuine, disciplinary limits as if they were capable of explaining every effect, phenomenon, and instance of sound. The laws of physics, assumed historical genealogies or religious practices shall thus sufficiently explain all cultural, individual, historical, and individual experiences with sound—at once. The situation, though, the context, the highly temporalized, spatialized, the culturalized, and corporealized character of a sensory experience is merely neglected by sonic transcendentalism as just some disturbing materiality and immanence with just so many annoyingly irrelevant details.

Claims in musical research that certain harmonies, instruments, rhythms, or syncopes do actually provide the same affective, metaphysical reactions in every humanoid alien no matter where she or he might live, might have been born, and what activities they might be indulging in, such claims perform sonic transcendentalism with a rather shocking degree of ahistorical and unlimited universalism. All in all, such musical mysticism, believing in an ahistoric and all-encompassing *sonic metaphysics*, necessarily fails in understanding or even simply acknowledging the functioning of a highly situated sonic experience—especially in the field of sound design. It might provide certain, though highly ambivalent, insights if applied to listeners and situations very close to listening situations that were structured by a belief in such sonic metaphysics. However, as soon as these very limited, historically excessively idiosyncratic situations of listening are left behind, as soon as a transcultural, a mobile and highly dynamized situation of vernacular culture is the field of research—even with differing sets of metaphysical or transcendental beliefs—these approaches fail or become at least rendered useless. I dance to these sounds; and in connecting various apps and my usual hardware appliances, I can craft a new stream of sound events, oozing out of a changing set of connected speakers. Sonic materialism as a method can truly be a remedy from such an overreach into transcendent and metaphysical territories. Materialism, sensualism, and a focus on the immanent and actual situation of listening and experiencing, this *radical empiricism* is one of the major approaches to open up the potential of, the possible encounters with and all the actual experiential and material qualities of sound. With such a materialist focus not a narrow but a wide range of theoretical assumptions, doubts, and threads of knowledge that are relevant to a *sonic experience* (Augoyard and Torgue 2005) can come into play. In this case, however, they do not overshadow and dominate the actual context of experience, the factual sonic experience: they are rooted and grounded in this experience.

Therefore, the analytical questions would be: How can one describe the potential of joy in sound and in sound design without evading into all too tempting though trivial desires for total excess and for legitimizing sonic transcendentalism? How can joy be an integral part of the everyday sonic experience? Can it be an integral part at all? These questions pose probably one of the biggest tasks for sound designers in all areas of design; and surely there is no simple and handy answer to apply in the process of designing sound following a sound briefing, in avoiding the dangers of annoying, harassing, even boring their listeners and users, turning their sonic consumers into nothing more than just consumer citizens. You put your headphones on to block out all the other colleagues' noises in this open plan office—or your kids' rambling and singing while you're in the home office. How can one by designing sound for a functional purpose evoke and stress in the listeners their personae of being *humanoid aliens*, being *sonic personae with a sensory corpus*? How can one avoid by designing sound for a functional purpose reducing them to simple clichés of *sonic consumers* and *consumer citizens*? Should you include all your kids and colleagues into your work? Could it be that a sole focus on certain *sonic affordances* in a design assignment or an underlying *sonic pattern language* will detour the sound design from its actual main goal? And what would this main goal then actually be? Who would dare to explicate that? What kind of excess would that result in—and would you, the sound designer, be hired for providing that?

If there is one major difference between sonic desires and articulations of sonic indulgence and joy and the dispositives in place to secure pervasive surveillance and a balanced economy, it is surely that there is a drift in sensory and sonic experience and *jouissance* that exceeds rather laconically strict tracking and accounting: a sensory drift of attractions and repulsions, of desires and indolence, of boredom and aggression, of erratic and idiosyncratic inclinations. Any institutionalized sound design with the quality of a dispositive surely rejects this drift. This sensory drift and one's urge for it is mainly the most extreme contrast to any aim for functionality, for effectivity, or guidance, it seems. I look outside, through the window, just following the blurred and vague movements of this ambient drone, oozing out of my speakers now. How could this drift then actually be useful at all in designing functional sounds? Does aiming at the impossible really help in transforming design processes? And how impossible is such a sonic drift really? Is it part of ordinary everyday life at all?

Situations in abundance

The sound of everyday life is never stable. The connections hum; sounds start too loud, to early, delayed or with unnecessary echo. The well-designed companywide sound logo covering kitchen appliances, personal computer gadgets, mobile phone services, and even automotive navigation systems might simply not work in all listening situations; the well-adjusted branding of an international financial services provider might be misinterpreted due to corruption scandals; the sound design for a food producing company in the sector of dairy products can sound outdated and strange because musical production styles develop rapidly and erratically: these are just some of the products whose sound designing process we encountered and took part in during our fieldwork. Sound jumps around, disappears, and returns. It bounces and bops, it annoys with repetition, it bores with unsurprising and tiresome non-glamour. It is always—seemingly—identical, endless, and without any erratic shock anymore in it. The same sounds of doors. The same sounds of alarm clocks. The same sounds of public transport. The same sounds of kitchen appliances. The same sounds of workplace machinery. The same sounds of entertainment gadgets. The same sounds of communication devices. The same sounds. The same sounds. The same, same, same, sounds sound. Same sounds. But are they? Are they the same indeed? Are they really only presenting mere repetition with no difference at all?

Sounds you or I recognize as completely identical might indeed be identical on the one hand in all properties of their electroacoustic or their mechanic generation. Yet, these sounds are, on the other hand, never actually the same. They are never actually articulated in a completely identical situation. The sound of its production propagates in ever changing ways and it is assimilated by humanoid aliens, by sonic consumers never in exactly the same way. You enter the subway as a Doppler-effect carries away the closing siren of another train. You log into this security system of the building you work in, and your rapid performance produces a glitch in the sounds, squeaking, repeating now on hold; a weird acid house loop on the spot. The audio guide of this exhibition space educates you while you simply continue listening to your favorite electronic music-record these days. The variations, the figurations, and constellations of sound generation and sound assimilation are *never* precisely the same. They transform, they change, they swivel and bop, they glitch, cuddle, stumble, stutter, and rush. These endlessly differentiated details of sound generation and assimilation as well as the endlessly differentiated constellations of creatures, substances, and artifacts therein, they constitute a sounding situation as a whole. A situation of sound is in flux by definition, "a temporal flux in which elements interpenetrate one another … an anonymous flux akin to the flows of minerals, biomass, and language" (Cox 2011: 148, 155).

This state of flux in sound generation and in sound assimilation though has been considered a major issue and a troubling problem for traditional research approaches of acoustical engineering, for room acoustics, and for the field of physical acoustic in general (Schulze 2018: 9–28, 88–94). Even more so, this state of constant and often thoroughly uncontrollable and in part unpredictable *sonic flux* (Cox 2011: 155; Cox 2018) seems to pose still today, in the 2010s, a major problem for many approaches to model and to conceptualize situations of listening and of sound propagation to their fullest extent. The efforts of acoustics since its advent in the nineteenth century tried and managed to a certain degree to sculpt an abstract, a formalized, and a diagrammatic core structure of sounding and listening. Yet, the questions at stake after roughly 150 years of research have shifted from the precise description and exact prognostics of a neatly limited, mainly physical, and in general abstractable and transferable phenomenon to the convincing interpretation of a highly specific, a thoroughly situated, a materially subtle, sensorially historicized, and biographically shaped phenomenon. Exactly those components that were regarded as irrelevant, accidental, and only marginal in the original conception of acoustics, all those entanglements and mixtures with areas outside of physics in its narrowest sense, they indeed have more and more become the really interesting, intriguing, disturbing, and highly problematic issues in sound research. These manifold and excessively volatile constituents are now characteristic for *the abundance of a situation*. Abundance is hard to research on; scarcity in contrast is much easier to assess (Schulze 2018: 88–94). Hence, exactly in this area of troubles, of disturbances and distortions the substance for another, differing approach to sound in general and to sound design in particular can be found: in the abundance of any situation related to sound. As in the words of psychologist and phenomenologist Eugene T. Gendlin (Schulze 2018: 145–150):

> Any situation, any bit of practice, implies much more than has ever been said. (Gendlin 1992: 201)

This diagnosis alters the fundamental relation between experience and language. With this axiom in mind any sensory and situated experience and its descriptions, in any kind of code, are attained with a different impact. This new and thoroughly reversed assessment of the status of language and codification then has obviously also a strong and lasting effect on any research and especially on developing cultural theories—be it on sound design. Whereas a transformation of sound theories and sound concepts through an enhancement via an anthropology of sound was discussed in an earlier volume (Schulze 2018), this volume will make an effort to expand the concept and theories around sound design, drawing from this earlier study. This expansion, hence, moves a currently predominant and hegemonic concept of sound and listening—in production, in the arts, in design, in signal transduction—that is structured by the idea of a scarcity in means, in material, in situations, and in communication into a concept that acknowledges the fundamental richness and abundance of any listening situation, of any sounding device and practice, of sonic bodies and of listening habits. The reduction to a scarce signal is being rejected and exchanged with an expansion of multiple affects, of extremist idiosyncrasies, and of volatile experiences.

In order to achieve such a transformation the researchers who worked on this project chose various approaches to implement this approach in their investigation. When the actual and abundant situation of listening and sounding is indeed the realm where meaning emerges out of a constellation of sonic encounters, then methods from the arsenal of *field research* and of *participatory observation* become obviously the major if not only feasible methods. In the case of sound studies these approaches are then expanded into the auditory forms of research such as *field recording* and *participatory listening* (Forsey 2010): from traditional ethnography emerges a sensory and a *sonic ethnography* (Pink 2009). The individual idiosyncrasies of present researchers become then, unsurprisingly, the instruments, the probes, and the media, of these inquiries. With these we hoped to get at least a certain hold of the sound practices, the corporeal listening practices, and the highly idiosyncratic experiences shaping and fueling the everyday situations of sonic labor. In contrast to a sole focus on philological interpretations or acoustic analysis of an auditory scene this major focus on individual experiences of a situation can open up the research process for all those idiosyncratic, corporeal, erratic, and sensory factors crucial in design. Still, an almost traditional *source critique* and *critical analysis* of the field research material remains a necessary and major task in research. The major difference in this case, though, is in fact that the material here is not being provided by institutions, actors, or groups of people who vanished or died long ago: this common assumption of deceased authors might seduce a researcher to assume they could interpret, reorganize, and newly frame this material in any way they like to, as they must not be too respectful with its authors. I object to this assumption. If the authors or protagonists are still alive, are still working on further cultural artifacts, are still developing their working styles, their aesthetic preferences, or their sensory idiosyncrasies, then the respect for their individual actions and backgrounds, their explicitly stated goals become crucial. A researcher who works *with* artists, performers, practitioners, or writers should never betray themselves, they could easily superimpose an interpretation on some object of research: one is then indeed working with, in a relation of symmetry if not reciprocity, feedback and response. In this mutual exchange, the *joy of research* can actually be found. A joy, however, that is truly different from the joy of the monologue and the violent misinterpretation—it is a joy of a dialogic if not polylogic exchange, a dynamic and plastic multiplicity in flux. An abundant situation of sonic experiences demands such a dynamic approach to research.

The joy of research

The research for this book was mainly done as part of the project *Functional Sounds: Cultural Theory and Design Theory of Non-Verbal Sounds of Communication*. This project was situated at the Humboldt-Universität zu Berlin and the Leuphana Universität Lüneburg over three years, between 2011 and 2014, when our research received substantial funding from the German Research Foundation DFG. In this period, our team of all in all five researchers, young scholars and experienced researchers, explored the everyday practices of freelance sound designers and of sound design departments mainly in Germany, in Switzerland and Austria, also in the UK, in Denmark, the United States, and selectively in Spain, France, Japan, and Sweden. Moreover, various prestudies, workshops, in situ-experiments, and periods of participatory observation as well as in-depth analysis of design theories, internal process documents, intermediary presentations, and preliminary sketches belonging to the field of sound design did take place outside of this core research time—spanning over a distance of almost ten years, from 2008 to 2017. The core team of Julia Krause, Carla J. Maier, and Holger Schulze—the collaborators of this book—was at times supported and expanded by various researchers and sound designers such as Georg Spehr and Max Schneider. The methodological and epistemological crucial praxis of interweaving field research with critical analysis—which has become a marker of the Sound Studies Lab since then—was strongly inspired by discussions and collaborations with and the works by Karin Bijsterveld (Bijsterveld 2008, 2018) and Jonathan Sterne (Sterne 2003a, 2011, 2012): two colleagues whose impressive research projects, explorative courage, and analytical rigor had a lot of influence on the design and the process of our research over the years. The results of our inquiries are now presented and displayed in this volume and its four distinct perspectives in four parts. The joy of research might be found in the results of the sixteen chapters of this book. We hope to grant hereby an insight into the complexity and diversity of this very field of practice and its current interpretations.

Subsequent to this general introductory part, we will explore in the following four chapters the actual and empirical situation of "Sonic Labor": a quick insight into the everyday tasks, the material representations, and the current state of sound design. In escorting various sound design projects, with interviews, and additional questionnaires we excavated the actual material of sonic labor as it is experienced by sound designers themselves: starting with the "Location and Apparatus" (Chapter 5) that are actually used to produce new sound designs; over the individual "Skills and Habits" (Chapter 6) a sound designer might have acquired or learned over time in his professional practice; the "Conflicts and Heuristics" (Chapter 7) which a sound designer encounters in his everyday work and which he or she develops over time; and finally the major characteristics—sometimes irritating if not unsettling—of "Portfolio and Presentations" (Chapter 8) dominant in this field of the creative industries. These four chapters present their argument though in quite an unconventional way. The argument and its proofs are presented mainly as *visual* documents, as graphics and as visualizations—sometimes with an ironic or sarcastic twist (and now and then a longer or shorter caption to guide your imagination and to propose unsuspected interpretations). This series of four chapters stands out and connects the sonic issues of our research with the visual, the spatial, the haptic, and the habitual in sonic labor. It guides the reader into the immanence, the actual work situations with their inner contradictions, erratic constituents, and surprising connections. You might browse and laugh, doubt and feel anger, be annoyed and perhaps even recognize your own design practice or design experiences on the subsequent pages of Chapters 5–8. The results of our fieldwork you can mainly find in these chapters that do not lay out a neatly polished and suspiciously consistent interpretation of the field we researched, but a collection of sometimes more isolated, sometimes more connected, but in all cases well-documented, situations of sonic labor. However, the verbal argument returns as the main actor of this book in the following third part, "Living With Sound," in Chapters 9–12, with discussions of the most intricate entanglements of current critiques in sound studies, in epistemology, in media studies, and in postcolonial studies based on our empirical studies on sound design and on additional field research. In this part we explore what happens as soon as conceived sounds are out in the wild, being domesticated by listeners, users, clients—all *doing things with sounds*. We start with scrutinizing the detailed effects that can be tracked in the semiotics of functional sounds as a way of "Signifying Sounds" (Chapter 9) before we move into specific sound practices in personal situations as a form of "Situated Signaling" (Chapter 10). From these fundamental explorations of everyday sonic experiences we will then move toward a kind of *sonic mediology*, following the work of cultural historian Régis Debray, discussing the processes of culturally "Transmitting Sounds" (Chapter 11). These investigations then come together in a discussion and a proposal to decolonize sound design under the title of "Transcultural Aurality" (Chapter 12). These two main parts, "Sonic Labor" and "Living With Sound" present, discuss, and expand our empirical material on sonic labor and sonic consumption in order to achieve a complex, complementary, and rich set of perspectives regarding the roles and functions and appearances of sound design in contemporary mediated and networked societies.

The last four chapters of this book, "Sound Works," propose an all-encompassing synthesis to all these scattered observations, surprising insights, erratic discoveries we as a research team encountered on our journey. It proposes a fourfold but coherent *cultural theory of sound design* derived from four major results of our research. This theory draws from the previous chapters and sections and presents an experimental, at times surely provoking if not disturbing and often willfully extremist interpretation of the past, the current, and the future state of sound design. These four final chapters discuss "The Silencing Dispositive" (Chapter 13) as a major prerequiste for any practice of controlling sound, defining functional listening, and securing the admission, presence, and correct delivery by consumer citizens; they explicate how "The Economy of Sound" (Chapter 14) actually works as sonic labor constitutes a sonic workforce stuck between ambition and exhaustion; the sonic consumers driving it therefore actually maintain "The Panacoustic Society" (Chapter 15), externally stabilized by the more and more visible forces of authoritarianism and their ubiquitous domestication of a pervasive sensology of surveillance. In the final chapter of this book "The Sound and the Situation" (Chapter 16) are presented and discussed as the actual constituents of sound design under the perspective of a decolonized, historicized, and a sensorially nonsegregated interpretation of sonic materialism: how to understand sounds, how to react to and how to propagate sounds, and—last but not least—how to interpret and thus to live with sounds. In this part our book connects predominantly our earlier observations, reflections, and insights in this book to the wider contemporary discussion in critical and in political theory, in social and in cultural theory.

This book, *Sound Works*, moves after a perspective on the general historical and theoretical overview in this part now to the perspective of a specific work situation: one precise location, where a sonic consumer, a protagonist of sonic labor, is listening to sounds and also working on sounds. What is your *private workbench*, how do you *stage your startup* company, on what *big mixing desk* do you present your work on? Let us join the designers in their studios, their open plan offices, freelance home studios, or niches crammed into a living room. We hope that the provocation that this book—and especially the next section—might represent can incite further development and further transformation in the field of sonic labor.

PART TWO

Sonic Labor:

Statements, Situations, and Cases

With illustrations by Julia Krause

Where and how do sound designers present their work? What kind of equipment do they use? What do their work places look like before, during, or after a production? Is there a specific sound design language? Over the course of three years between 2011 and 2014—with preliminary and additional studies in the three years before and after—these and many more questions were explored by our research team. Participant observation, analysis of design artifacts, interviews, workshops, and a general as well as a focused questionnaire were the tools that helped us excavating the empirical material for this book. A total number of seventeen sound design projects in eight European countries and the United States were part of our research. The main results of these qualitative ethnographic studies can be found in detail in this section of the book called "Sonic Labor."

The field of sound design is still comparably small. Many of the actors, protagonists, clients and professionals may know of each other or are very likely to meet and collaborate in the near future. Hence, it was crucial to anonymize the designers, the collaborators, the clients, and also most of the projects analyzed and scrutinized in our book. This decision is the consequence of our wish to respect the professional integrity of all actors and their disclosure agreements. We wish to focus on the factual phenomena, problems, and inner contradictions of sonic labor instead of exposing highly idiosyncratic flaws, mistakes, or awkward bêtises as performed by professionals, companies, or projects.

The images, graphics, visualizations, and individual image captions were crafted and selected by our team with the intention to find a striking representation of the major insights of our research in this book. Hence, the material used in these illustrations and collages refer to research on sound design projects done by Holger Schulze and Julia Krause as well as some design projects encountered and analyzed by Max Schneider and Georg Spehr. A smaller but still significant number have been produced and provided by sound designers, clients, and professionals of sonic labor that we have been in conversation with; in these cases, their (verbal or visual) contributions are obviously given credit. The visual representations and collages are designed by Julia Krause: she took up insights, results, evaluations, and aspects of the field of sonic labor as it presented itself to us in our research—through participant observation as well as through analyzing selected sound design products—or as discussed in the interviews we conducted.

The captions are written by Holger Schulze in collaboration with Julia Krause, a collaboration that included ongoing discussions of our research results and forms of representation. The captions provide sometimes suggestive, ironic, even sarcastic, poetic, narrative, or surprising if not erratic interpretations and accounts that are grounded in the material we gathered from the interviews with sound designers and people working in sonic labor. They were also rewritten and refined in relation to the visual representations by Julia Krause with the aim to put into a dialogue both the verbal and the visual dimensions of our analysis and to enhance their suggestive and thought-provoking character. These captions are not direct quotes, but they embody and stage major positions, statements, or recurring explications made by several of our interview partners. In the rare cases in which captions resemble precise quotes from specific interviews, texts, or questionnaires these quotes are referenced.

The scenery of sonic labor as depicted on these pages, is not a mere illustration that accompanies our writing. It is a substantiated result of our research and reflects an integral part of our analytical work and our interdisciplinary perspectives of anthropology, sound studies, cultural studies, performance and visual arts. This outcome can be regarded as an effort to visually expand the analytical, the historical, and the political dimensions that are also very much present in the three written parts of this book. This display, though, is—needless to say—not comprehensive and not exhaustive. It was not our aim to represent the totality of all sound design processes, or all design cultures, in history, present, and future. The research behind this book has a specific angle, it is limited in time (2011–2014) and space (mainly western and northern Europe with selected sidesteps into other countries), as such it also is limited in the sound design culture that is represented: a still heavily male dominated culture. These limitations, however, allow us to analyze approaches, practices, and habits and their social and cultural functions. Our study provides therefore a series of casuistic and representative insights into the everyday practices and experiences of sonic labor. The research results on sonic labor as presented, synthesized, and interpreted on the following pages, provide hereby crucial interpretations and questions that will be taken up in the subsequent parts of the book that are geared toward a cultural theory of sound design.

5

Location and Apparatus

Staging the startup

FIGURE 5.1 *What do these buildings do? How do you work in them? How can you design sounds within these walls?*

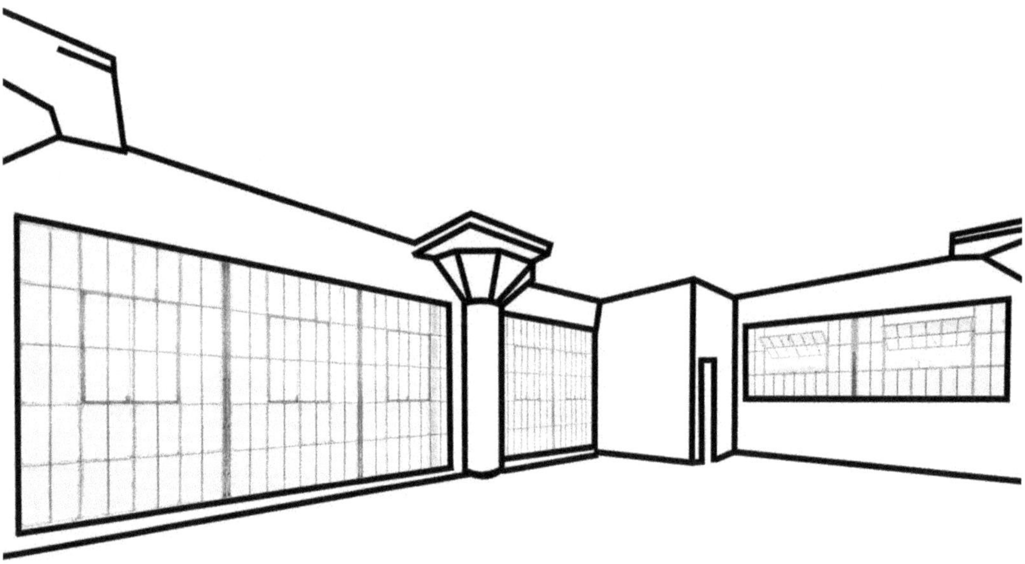

FIGURE 5.2 *This emptiness is promising. You can still smell the metal, the grease, and all of the chemicals used in this printing facility based here until* ca. *1978. Since then, all the dark machines have vanished, now we indulge in huge windows and sleek, shiny tablets. In our breaks we play table tennis or hang out in the snuzzle area.*

FIGURE 5.3 *I like to work at home. It's cosy and everything is nearby. Okay, I work only with headphones, so I don't disturb my neighbors; but when I have a great idea in the middle of the night, it's just two steps to my studio. I produce my dream sounds here (however, they often end up being the darlings that are killed later in the process).*

FIGURE 5.4 *I am on the road. It was not my personal wish, actually, but it just happened. The possibilities of being a nomad are great. A lot of people are doing it. I am flexible, Wi-Fi is everywhere and I am an addict of freshly brewed coffee and a handmade vegan cookie. The spectacular view on this pulsing metropolis triggers my creativity. And sometimes I feel really like one of these guys in a Netflix-series.*

FIGURE 5.5 *I prefer the good old way of doing things. You know, it's like working in another universe, in another time, and another space. The wood and the carpet exude a hint of dust, but it also calms you down. Actually, you can breathe in the* quality *and the* knowledge. *And this* silence. *Even the clock in the background is ticking so quietly. I get shivers.*

Private workbench

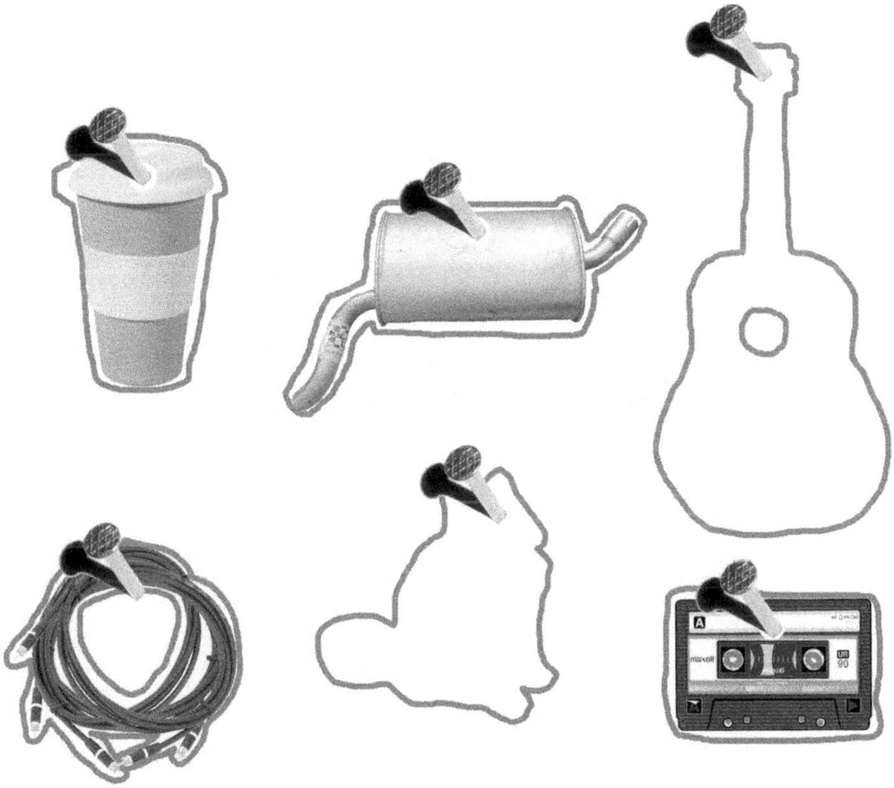

FIGURE 5.6 *What do you need in order to work on sound? What tools and instruments make you the virtuoso you are?*

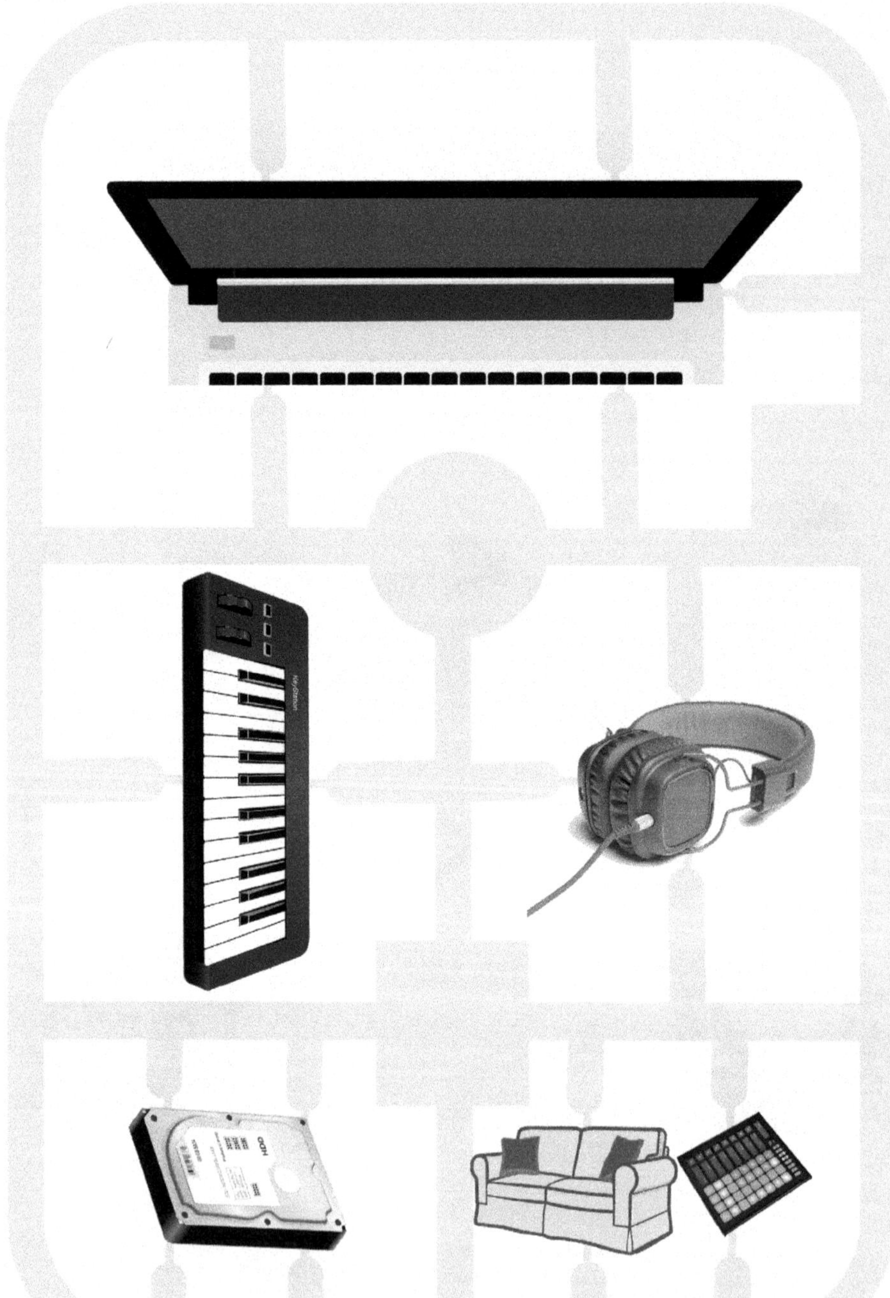

FIGURE 5.7 *This is your stuff, at home. Sometimes, a new device is added to your set-up, now and then a new computer or an update of your audio software; but in the end, this is all you need. Collected and arranged in a cosy spot, but not as inflexible as it might seem. Now and then you put it in a suitcase, and you can rearrange it on every stage, in every backstage slot, or budget hotel room. Home is everywhere.*

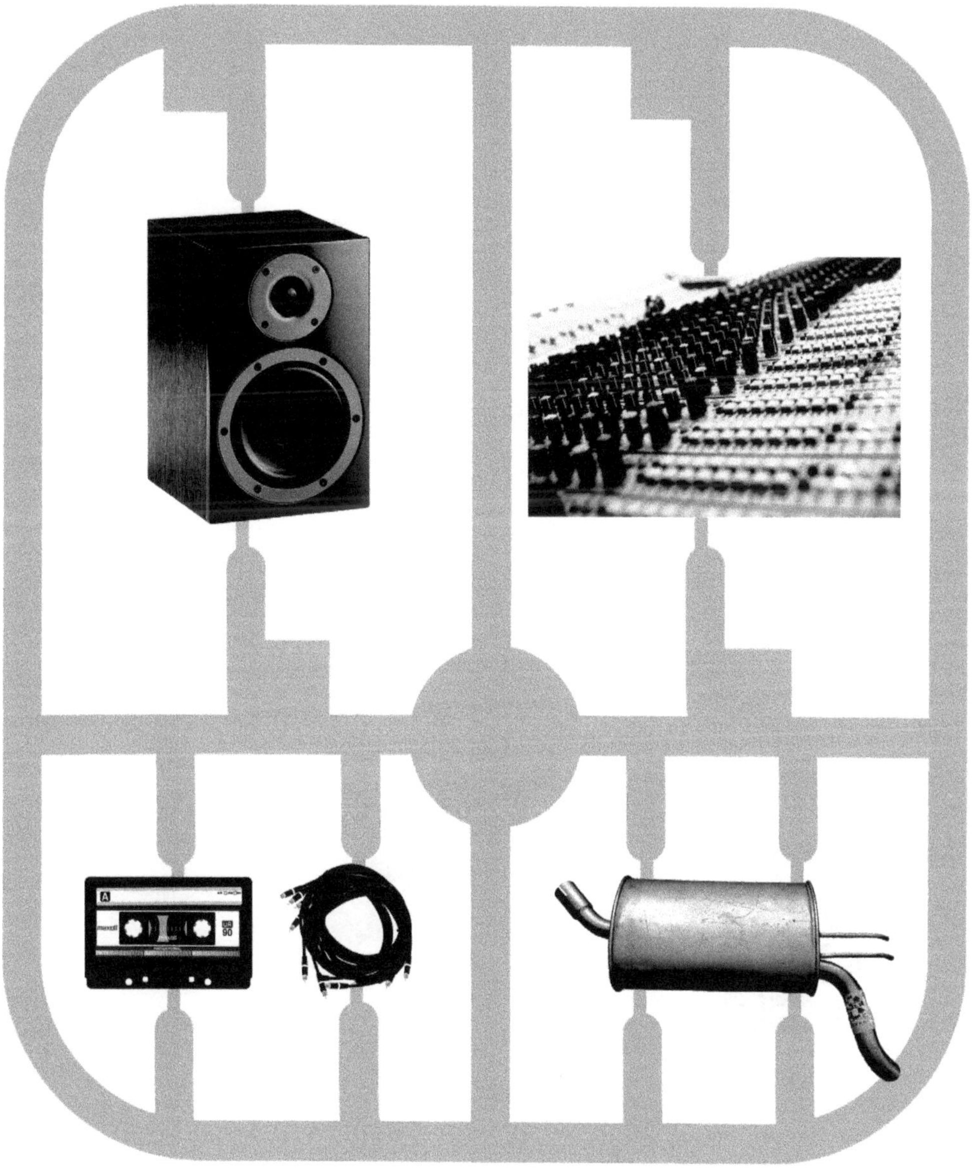

FIGURE 5.8 *Your studio is a precious artwork you built over years. Nothing is flexible. It should not be. For your work, exactly this constellation of tools provides the fine-tuned machine you need. Only rarely a new interface, new patches or musical instruments are included. But you stay reluctant. Never change a winning team.*

FIGURE 5.9 *You need to be hyperflexible, always on your heels. Hence, you produce the biggest amount of trash from coffee cups to napkins, boxes with donuts or ramen. You need to wait for the Wi-Fi to be working properly or just for a slot with enough power supplies in this coworking space. Concentration is a challenge here. Theoretically, though, it's a nice concept.*

FIGURE 5.10 *This is personal, strictly personal. The selection of objects might look weird, others come by and wonder why? But these things, mixers, generators, old toys, they make you happy with their sounds you find nowhere else. Sometimes, you tinker with them.*

Suites of production

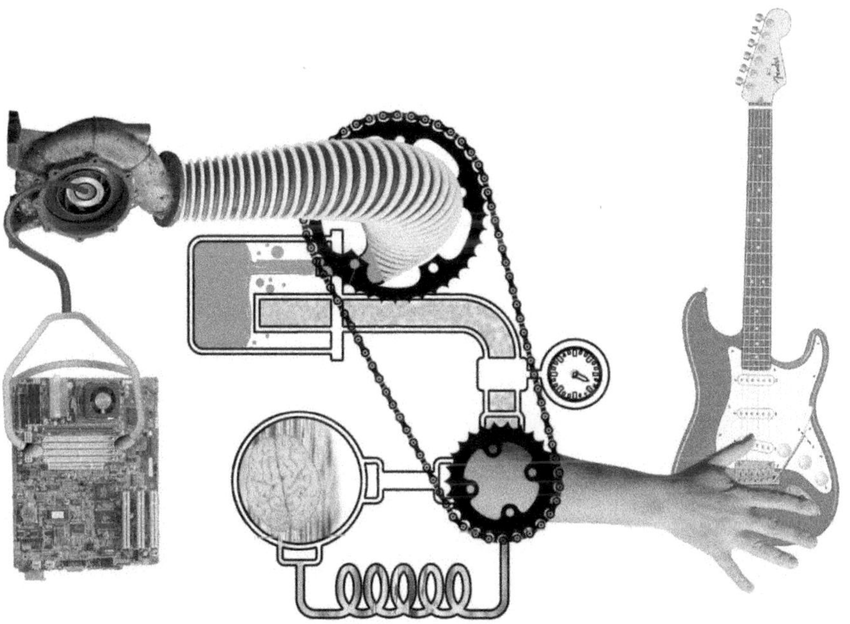

FIGURE 5.11 *"All that works is the sonic plus the machine that you're building." (Eshun 1998: 188f.)*

FIGURE 5.12 The Passion of Musicking all the Time.

I prefer to work before lunch.
freelance sounddesigner

We need more courage for complexity.
head of a sounddesign agency

Concentration?
Not longer than two hours,
the rest of the day is just
doing random things.
freelance sounddesigner

There is still
a superiority of
the monotonous.
permanently employed
sounddesigner

And therefore,
we sometimes ask ourselves:
How does a square sound like?
head of a (sound)design agency

Aesthetics-wise I feel no restrictions -
even rubbish sounds can be fun.
It's all part of the craft.
freelance sounddesigner

Silence is the ideal. Not only in my office. But when
everything is quiet, the whole building - that's perfect.
freelance sounddesigner

I don't care what other designers are doing.
permanently employed sounddesigner

FIGURE 5.13 *The Joy of Focusing: The Joy of Doing the Impossible.*

Tendentially, we are asked to work on a project at the last minute.
sounddesigner

It's would be supernice, if you quickly just to do this... or that...
client

email ping-pong:
I send some stuff and they answer.
sounddesigner

There is no awareness of the process. Time and money are the key principles. And we - the designers - smooth it out.
sounddesigner

O sorry, I just forgot, but could you...
client

There is no extra time, no latitude. We often receive requests for changes late at night that are expected to be done by the next morning.
sounddesigner

It seems to me that this short term working mode got more extreme.
head of a sounddesign agency

We do have a extrem high portion of costs for aquise and marketing.
head of a sounddesign agency

I have a drawer full of ideas, sketches and proposals. But you can't use this stuff again for another project. Even when you think you could reuse something, at the end you create something completely new.
sounddesigner

FIGURE 5.14 The Next Milestone was Yesterday (and the Briefing Changed yet again).

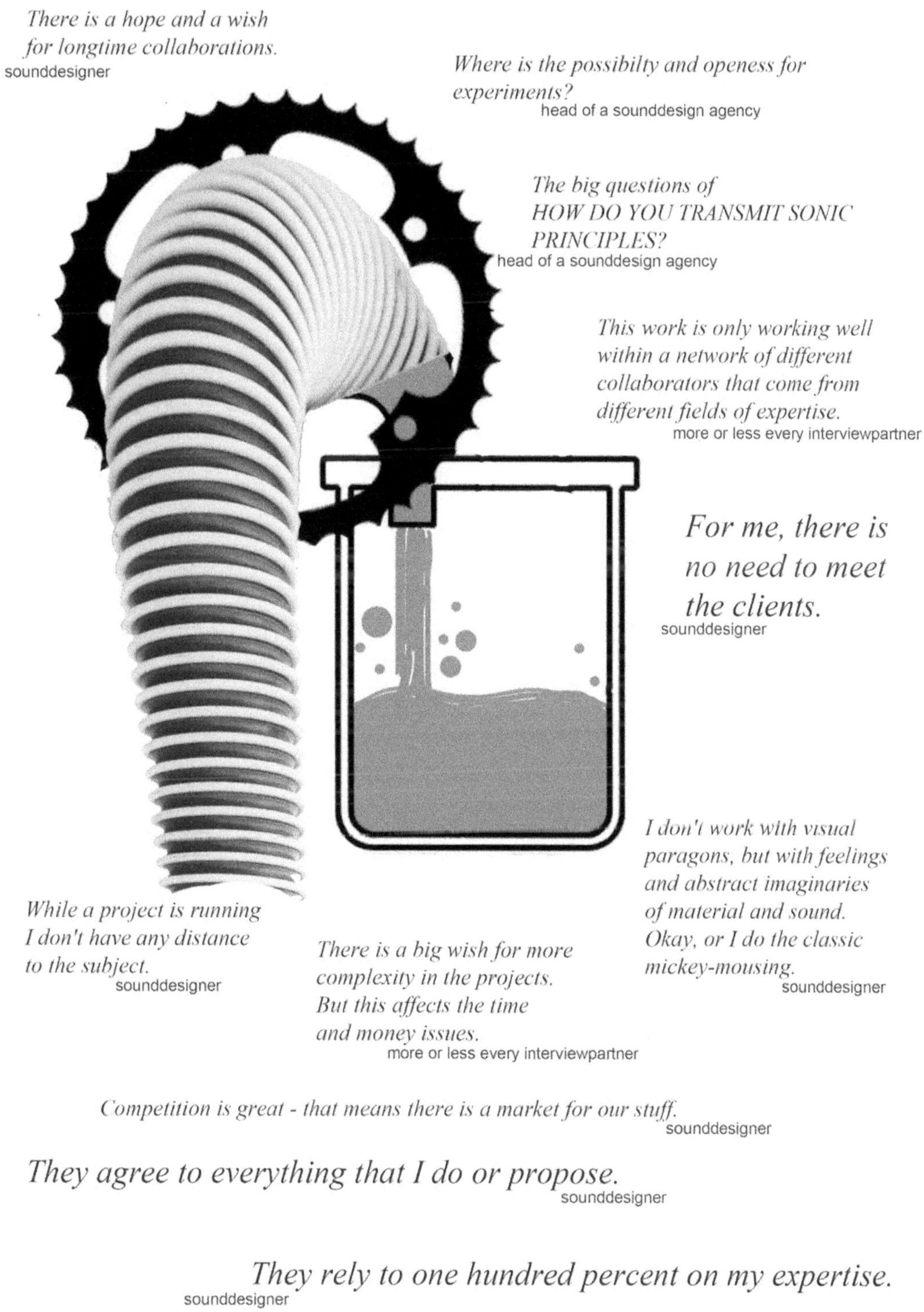

FIGURE 5.15 From Experimental Expertise to Implementing a Design.

The big mixing desk

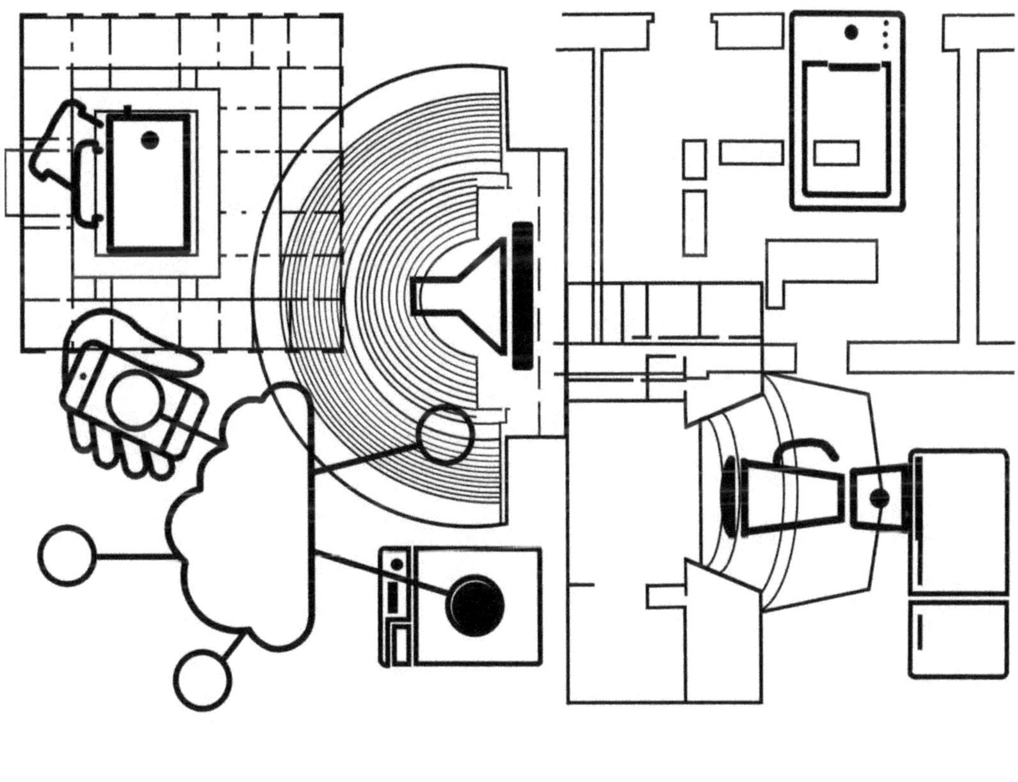

FIGURE 5.16 *How does all this work actually materialize? What are the objects, the settings, the institutions, the power structures you work for, actually?*

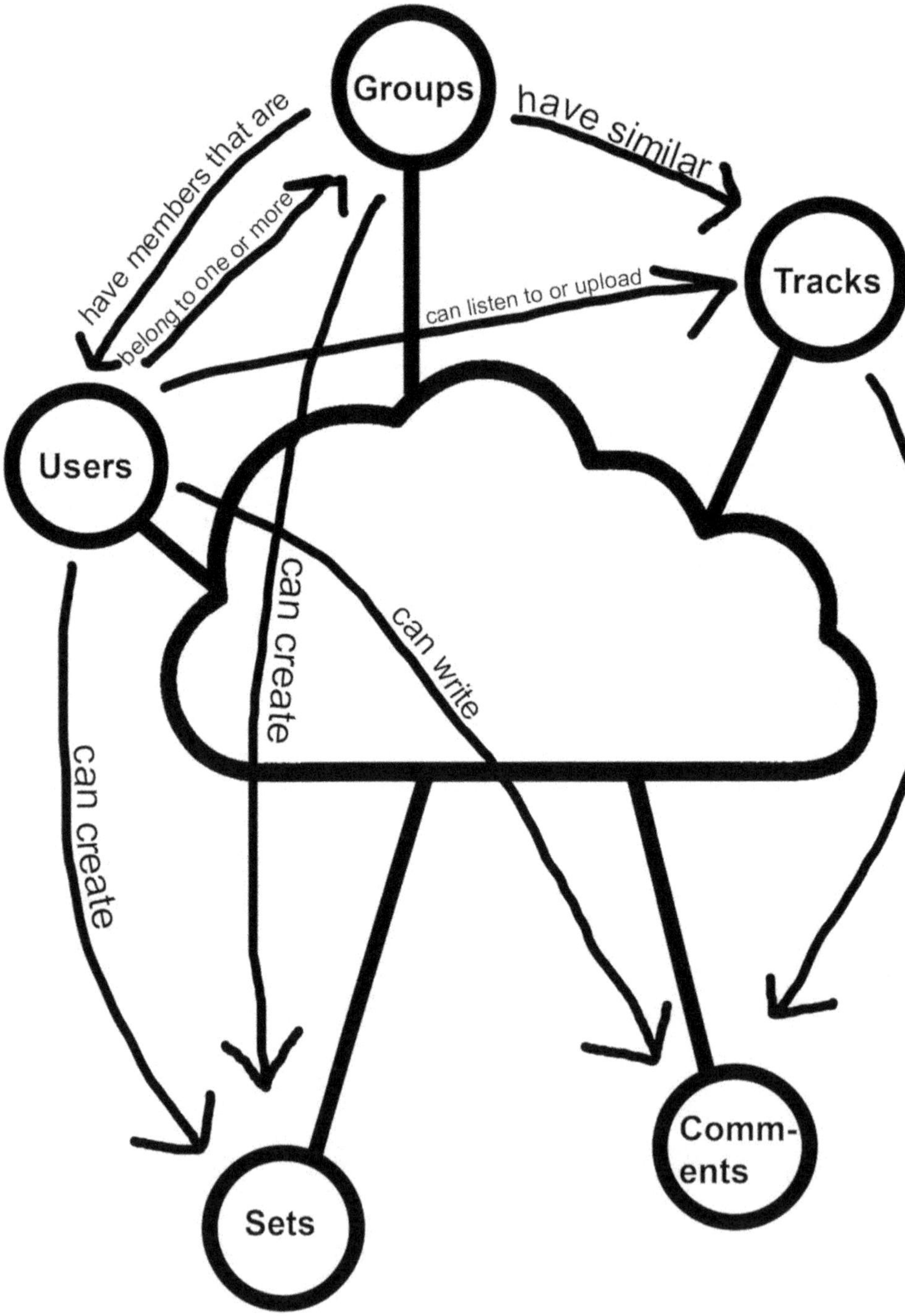

FIGURE 5.17 *Working in and for the cloud is quick: contacting our community there allows for rapid responses. We try to be quicker than anyone else, always.*

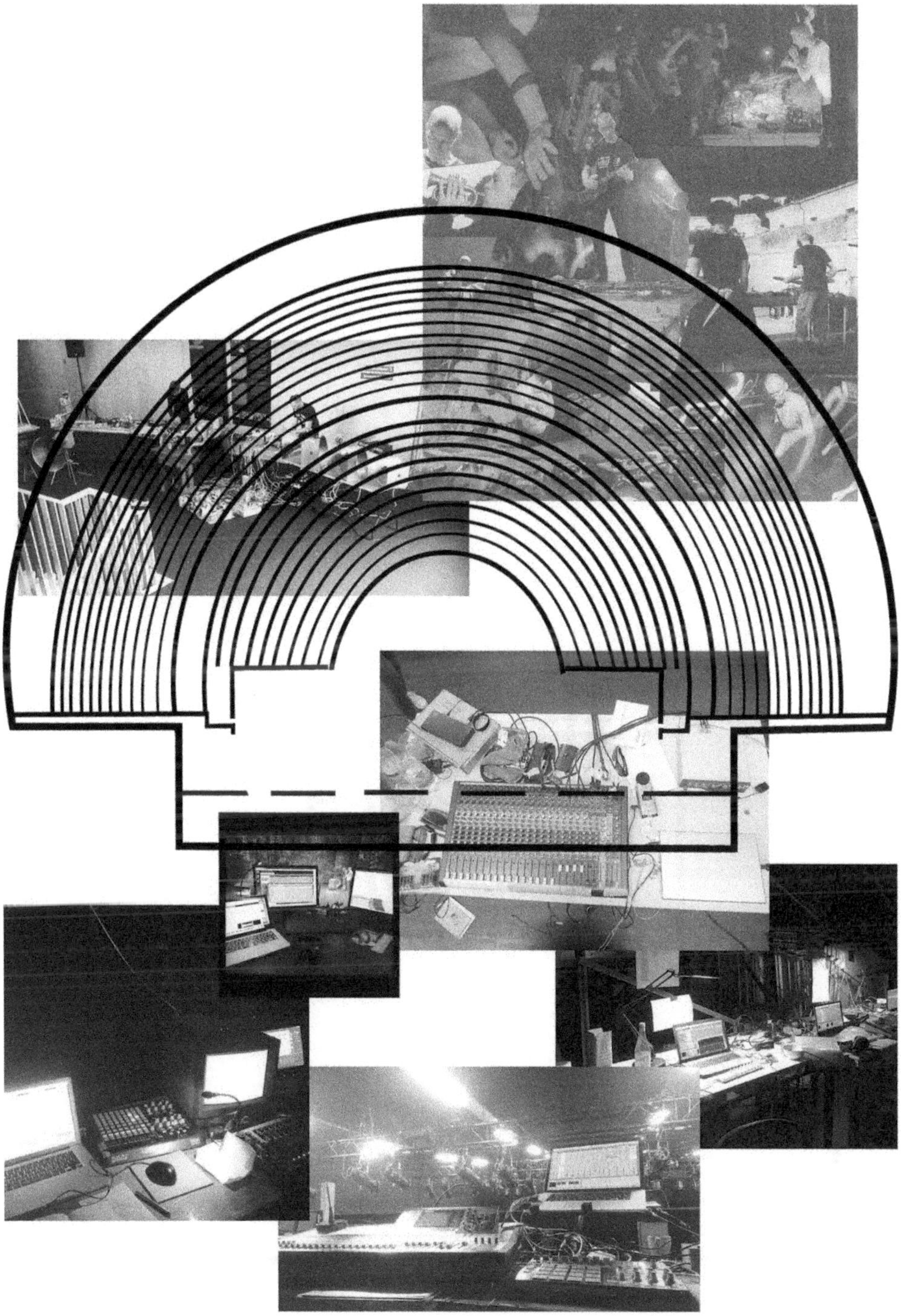

FIGURE 5.18 *A theater or a concert venue of a more traditional kind provide us with regulations and requests that seem often archaic now. As a sound designer, I am in close touch here to the audience—as well as to the light designer, stage designer, choreographer, the makeup artist. It is a constant, not rarely conflictual, conversation with many voices speaking up, most of the time. (Picture references (clockwise): Guido Möbius, Katharina Pelosi, Julia Krause, Matthias Meppelink, Julia Krause, Sebastian Waschulewski, Wessel Westerveld, and Yuri Landman.)*

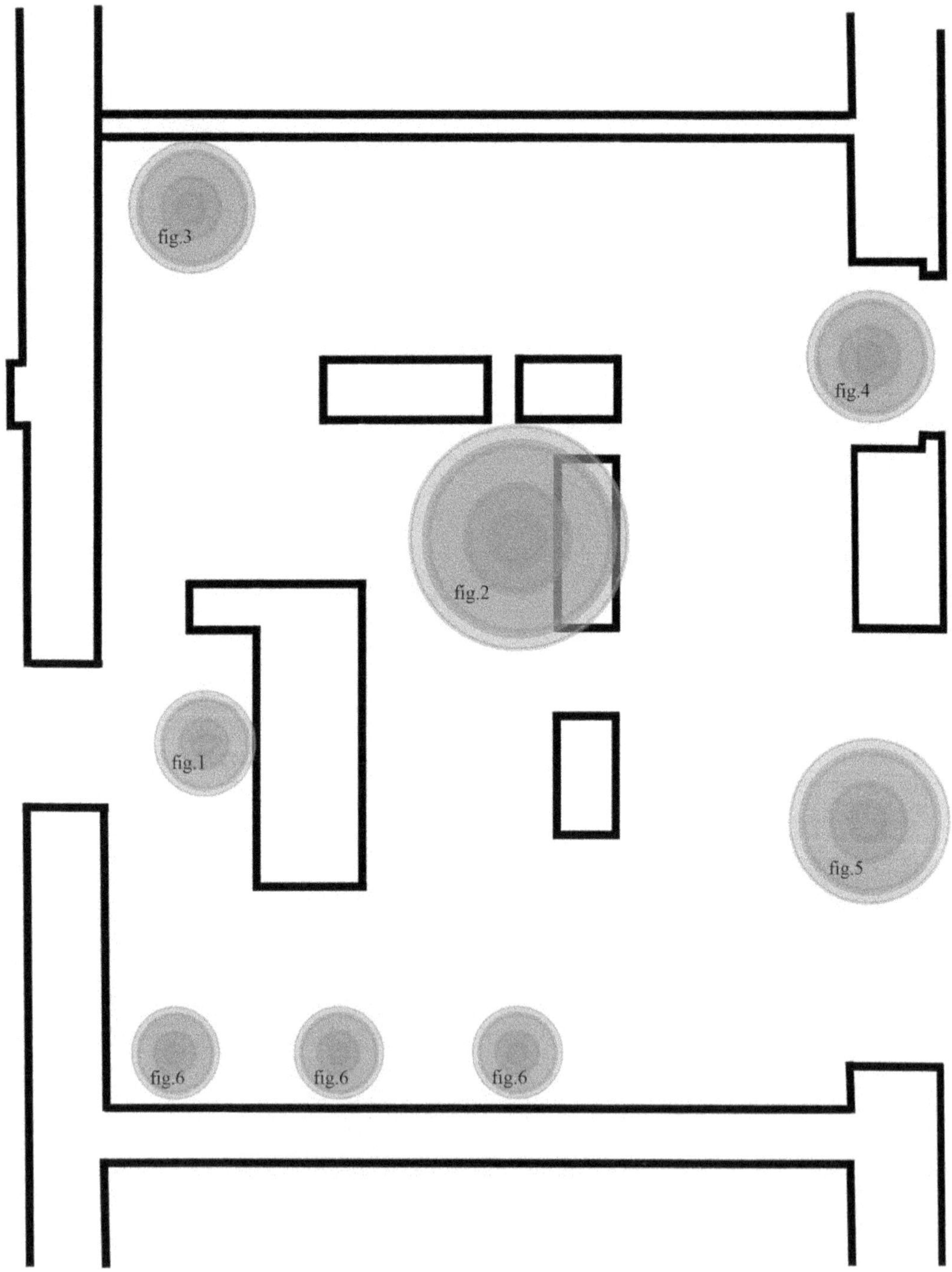

FIGURE 5.19 *A flagship store is a delicate affair. Those precious gems, new products, and noble visual logos. Yet, the so-called corporate identity is often only visual. You can invent and convince your client of a so-called* sound identity. *But mainly it needs to work at the point of sale. Is the* sonic consumer *satisfied? (*cf. *Chapter 1, section "The sonic consumer.")*

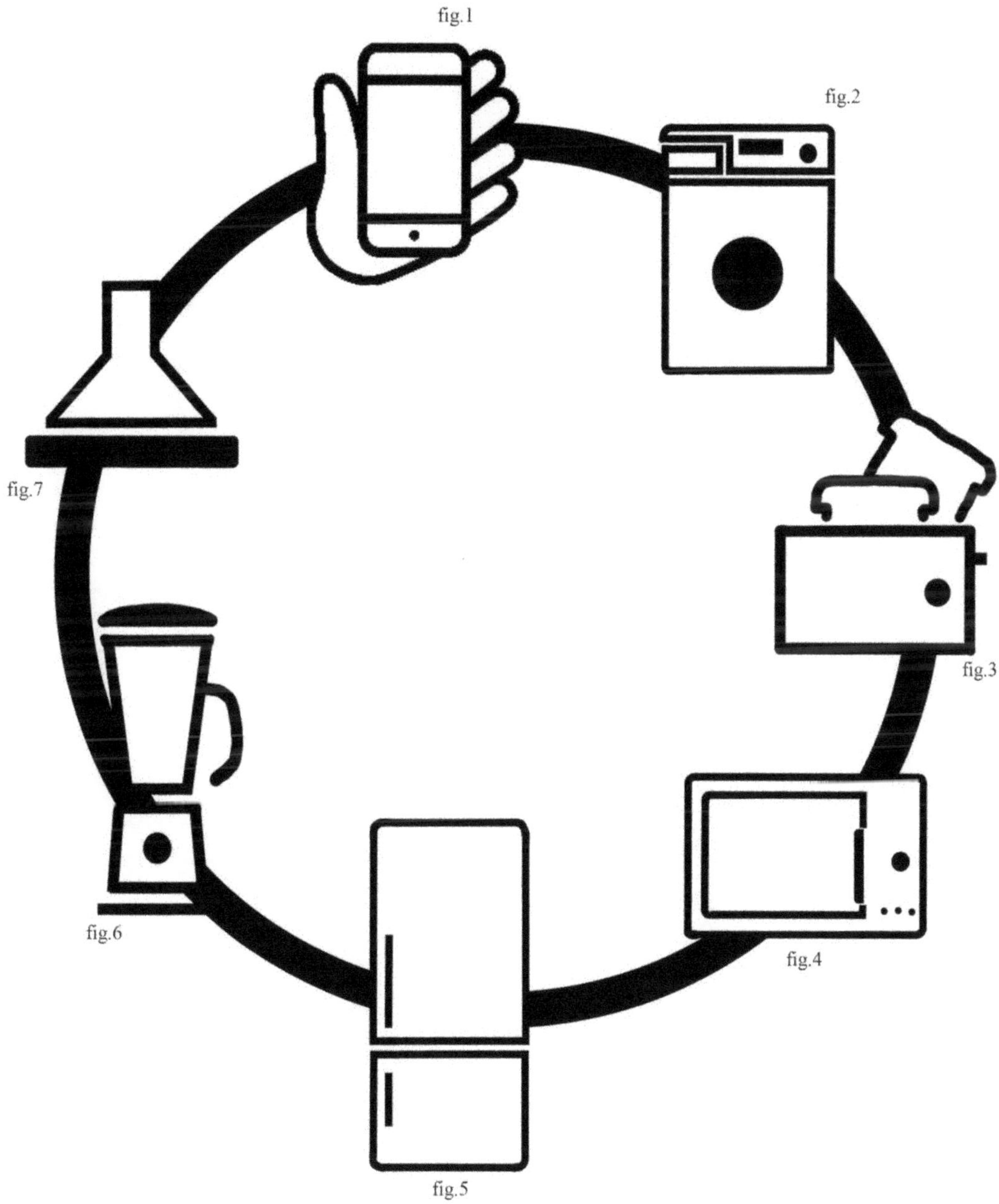

FIGURE 5.20 *Functional sounds need to fit the device, its use, its users, all third-party applications, and all the hardware; and the corporate identity. But do these* sonic patterns *really work? Or do other devices project louder and more intriguing sounds to the* consumer citizen? *(cf. Chapter 3.)*

6

Skills and Habits

Sonic imagery

FIGURE 6.1 *How do you explain a sound? What imagery to use? What gestures to make? What stories to tell?*

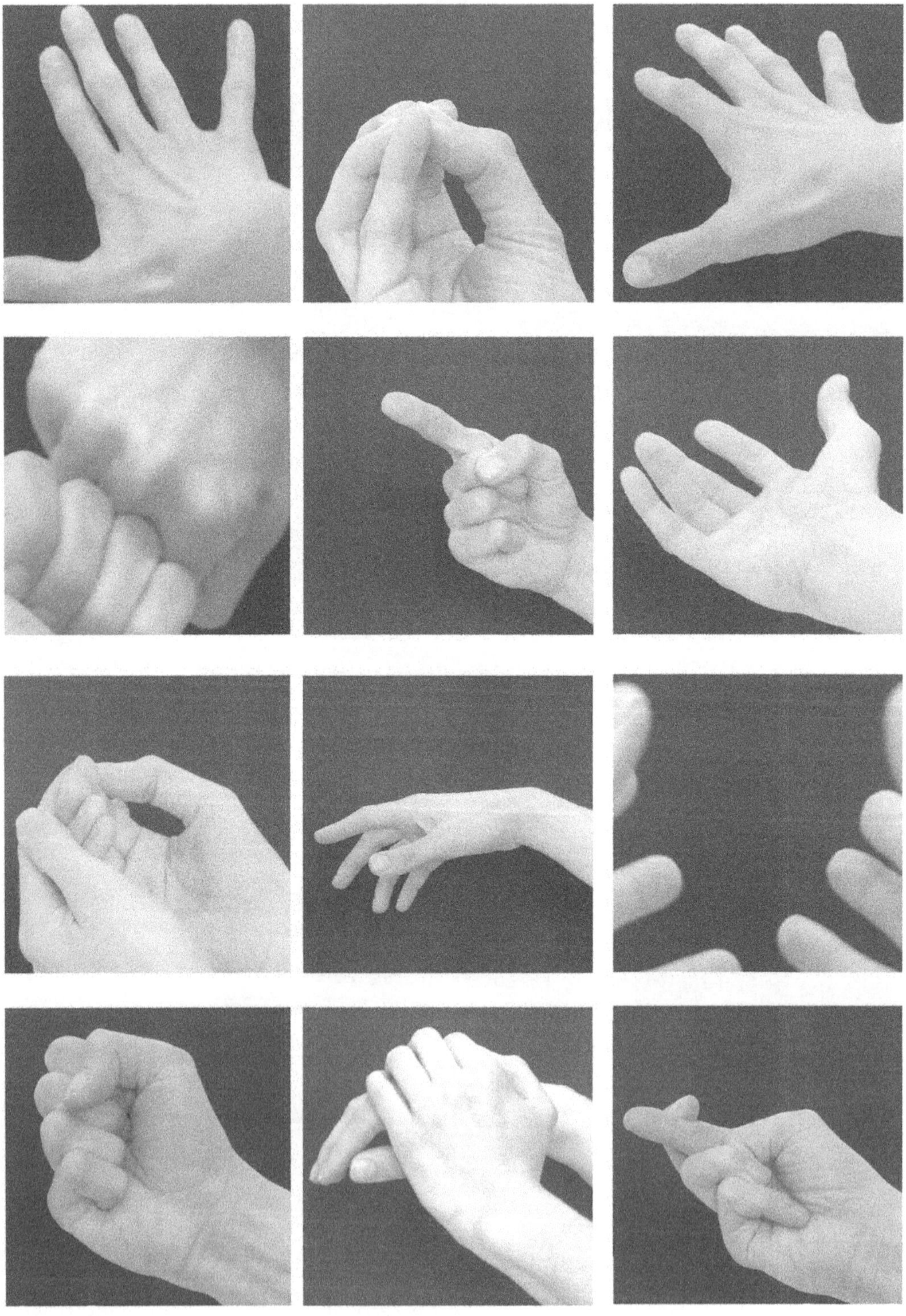

FIGURE 6.2 *Nine words to describe sound: which ones would you choose?*

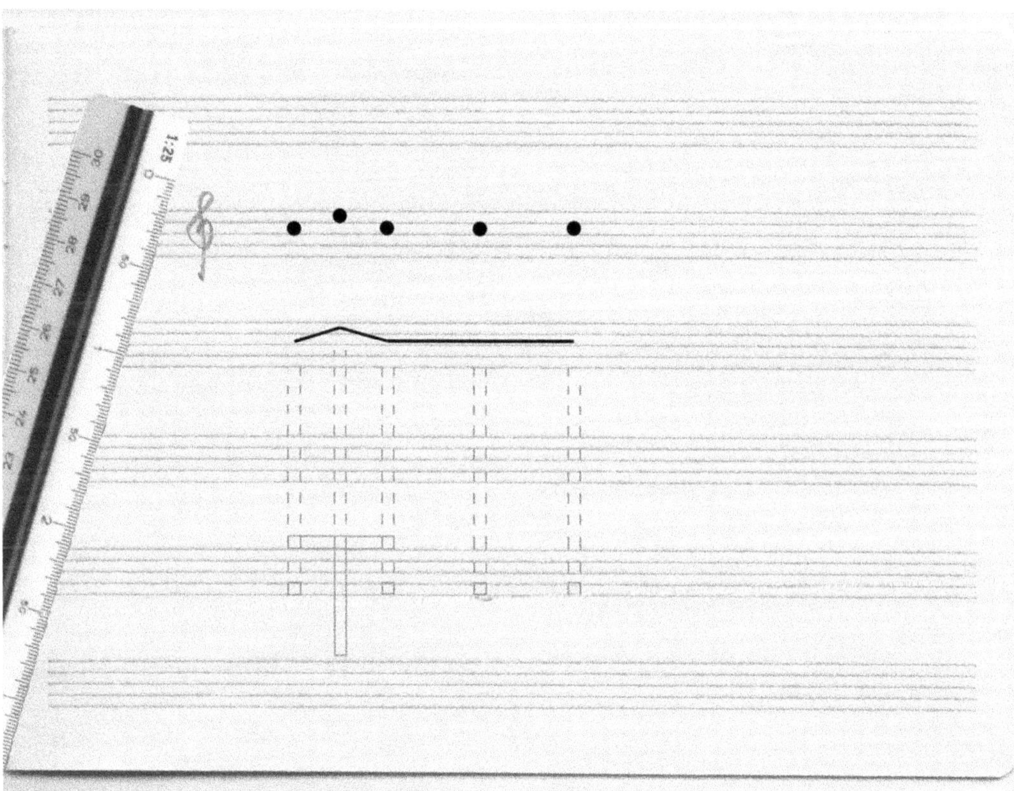

FIGURE 6.3 *Five tones to sonify a visual logo: how would you do it?*

FIGURE 6.4 *One scenic landscape to visualize your ideal soundscape: do you imagine what sounds?*

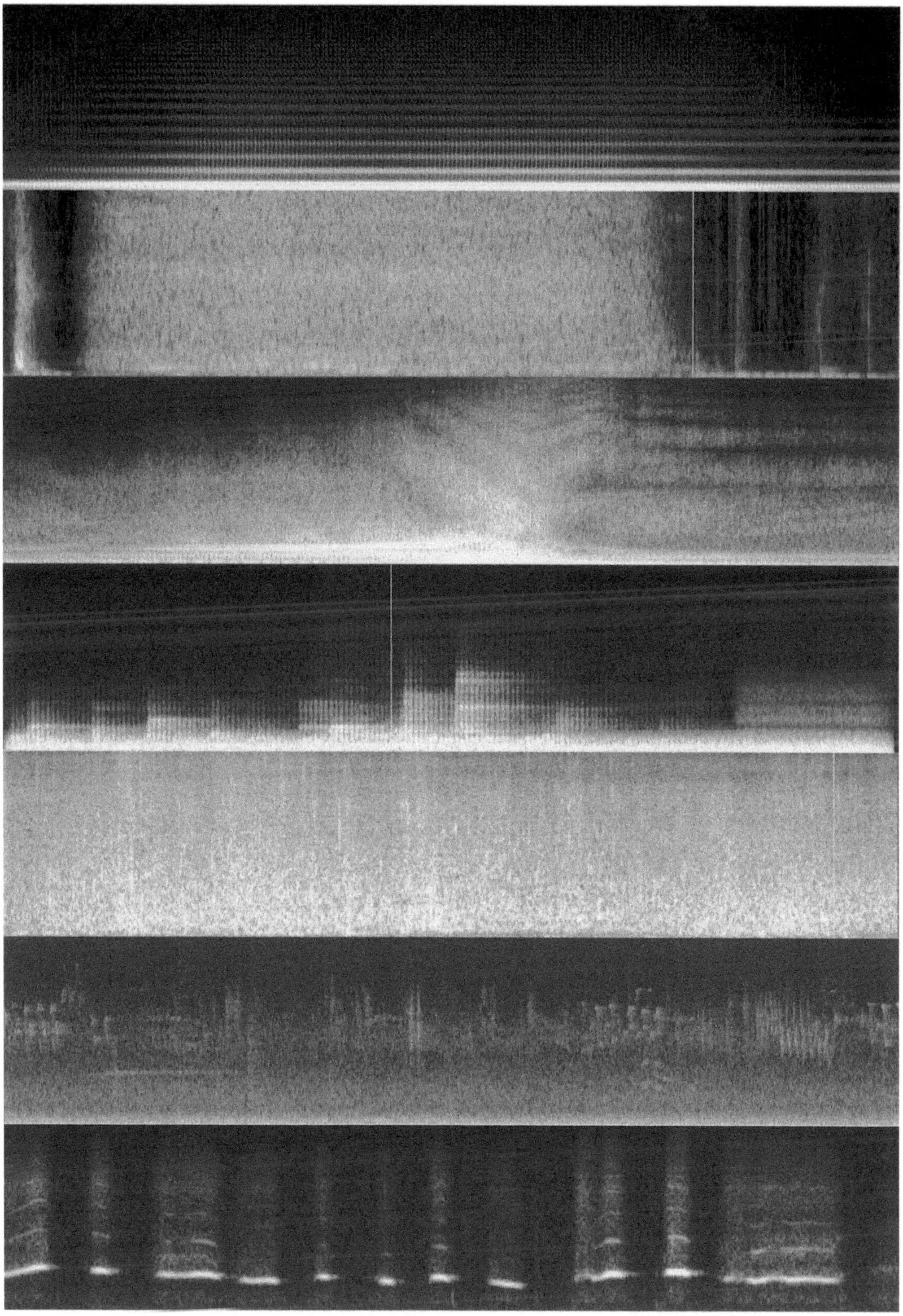

FIGURE 6.5 *Seven sonograms—take a guess what it is? Perhaps a radar? A balloon being blown up? Start of an aircraft? Space (as we think it should sound)? Waves on a tropical beach? A park in Paris? A whistling man?*

Make it fantasy!

FIGURE 6.6 *What private lingo do you use when designing sound? What internal jokes or weird, made-up words help you to find this specific sound quality you have in mind?*

FIGURE 6.7 THE JOE VOICE: *a rough and rusty style of singing, a rock voice; think:* CSI: Miami *opening theme.*

FIGURE 6.8 LOW HANGING FRUITS: *it is so simple, so apt, so really, really ready to catch—like a fruit that is very ripe, overripe. You don't have to stretch at all: they just fall to the soil.*

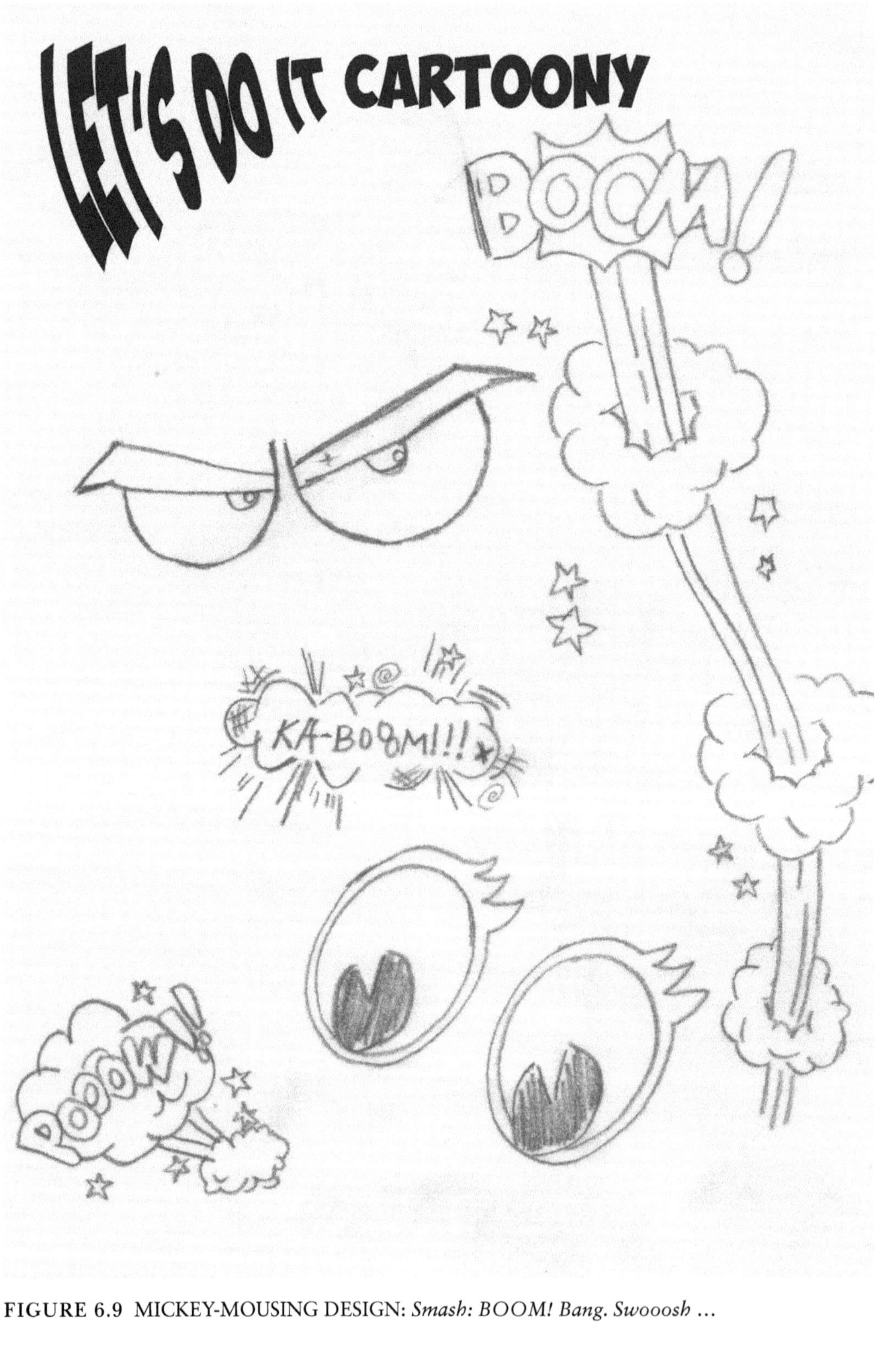

FIGURE 6.9 MICKEY-MOUSING DESIGN: *Smash: BOOM! Bang. Swooosh …*

FIGURE 6.10 HEROES: *a little more* Enoish*? More* Morricone*? Do we need some* Zimmer *here? An artist's name that stands for a whole spectrum of styles. Use it. Will you be just a copycat or a remixing genius? And why do you cite only males? Why not Lora Hirschberg or Anna Behlmer?*

New ear

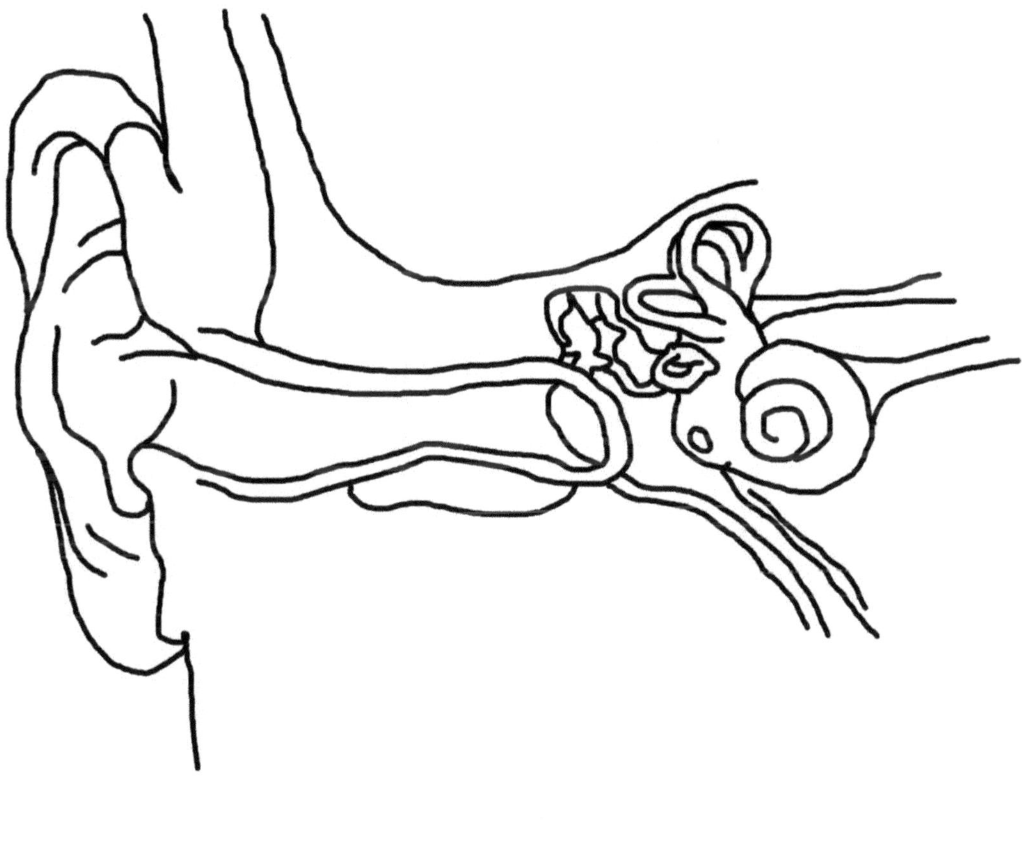

FIGURE 6.11 *How can you still carefully and meticulously listen to the sound you just designed? What advice do you have for younger designers to do so?*

FIGURE 6.12 *Go out on the street, with no devices. Think John Cage or Max Neuhaus. How are the window fronts reflecting the sound? How do the voices of passers-by alter when approaching and moving away? Something surprising here? Perhaps a hidden symphony, played by the car horns?*

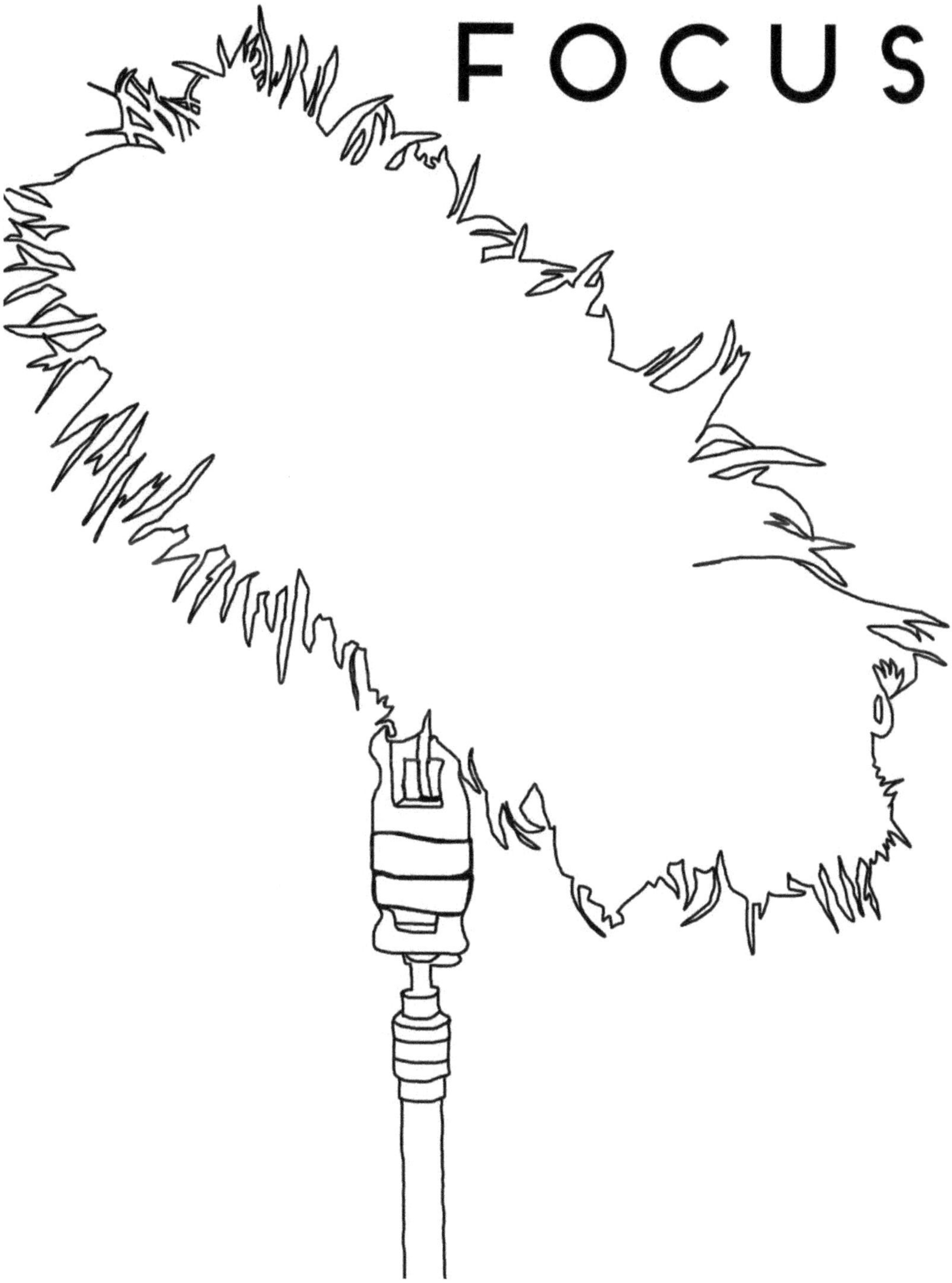

FIGURE 6.13 *Again, just go out, no matter where to—the woods, a street corner, your apartment, the library. But take your microphone and recording device with you. Listen like a recording membrane.*

FIGURE 6.14 *Go out and travel, all around the globe. Teleport yourself up into another cultural environment. Or no, better,* don't *teleport but take the long way, by train, by foot. Indulge in the tiniest differences. Meals, drinks, street signs, coats, jewelry, gender relations, small-talk topics, dance music.*

FIGURE 6.15 *Go. Experiment. Challenge your skills. Do something hard, difficult, strange, maybe appalling to you—for the first time (and don't think too much about the esoteric undertone in all these recommendations).*

The client whisperer

FIGURE 6.16 *How do you convince a client? And how does a client respond to your proposals?*

Message

To

From Cc Bcc

Subject

it should be a little bit more disconcerting
more general
something that tells you: you have to stop
at the end more open
like opening a window
a good feeling
save, strong and powerful
more valuable, please
it should be lighter
the ding-ding in the middle is too loud
more layers, more playful
even bigger
I don‘t hear a key moment
what is this at the beginning?
do you think that sounds like us?
more floaty, not so direct
like silence, but better
this is too definite
in the middle, something should drop down
could we work on the tension build-up?
it‘s too short
like a wake up my love call
this is making me happy
it‘s too obvious

FIGURE 6.17 *A brief feedback mail by the client can create some clarification. But more often than not it creates a lot of confusion.*

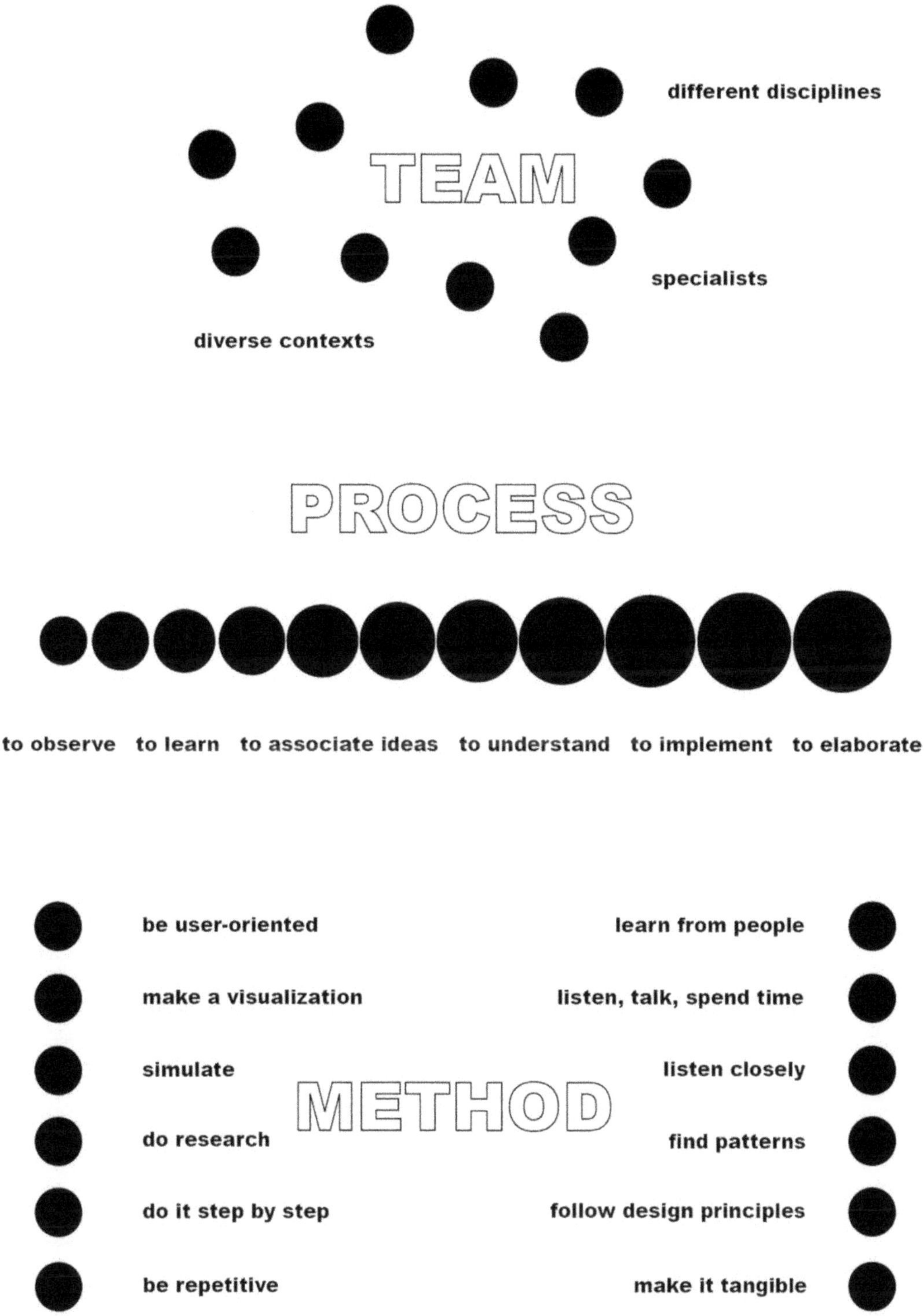

FIGURE 6.18 *With your flashy presentation you introduce everyone to your personal and professional method:* The Method: The Design Method. *The one and only, of course. There is no other.*

notes

FIGURE 6.19 *Let's make them hear. Open their minds for the issues of sound design by encouraging them to make their own experiences.* Not all your thoughts, doubts, and feelings! They are our material.

FIGURE 6.20 *You make the world a better place—by sound. You promise* sonic transcendence, *a metaphysics of listening, an audiopietism one could say. But you do so on the biggest stages, largest conferences, the most clicked streaming or video sites.*

7

Conflicts and Heuristics

Ephemeral lie

FIGURE 7.1 *Do you think sound is ephemeral and light and weightless? How silent is space anyway? The actual* sonic materialities *prove all fantasies of celestal, immaterial sound wrong.*

FIGURE 7.2 *Frying bacon sounds like it's raining down; cut a cabbage and it sounds like a punch with the fist; your hands in wet spaghetti sound like washing someone's hair; whirl a towel and it sounds like the flapping of a bird's wings. Unfolding cellophane wrap sounds like a burning cigarette, and a bicycle spoke played with a cello bow creates a horror atmosphere.*

FIGURE 7.3 *The* Nokia *ringtone, actually based on the piece* Gran Vals *by Francisco Tárrega from 1902, is still today one of the most played tunes around the globe (Gopinath 2013).*

FIGURE 7.4 *Any sound* could *be found in the databases of the World Wide Web. But will you ever find it again?*

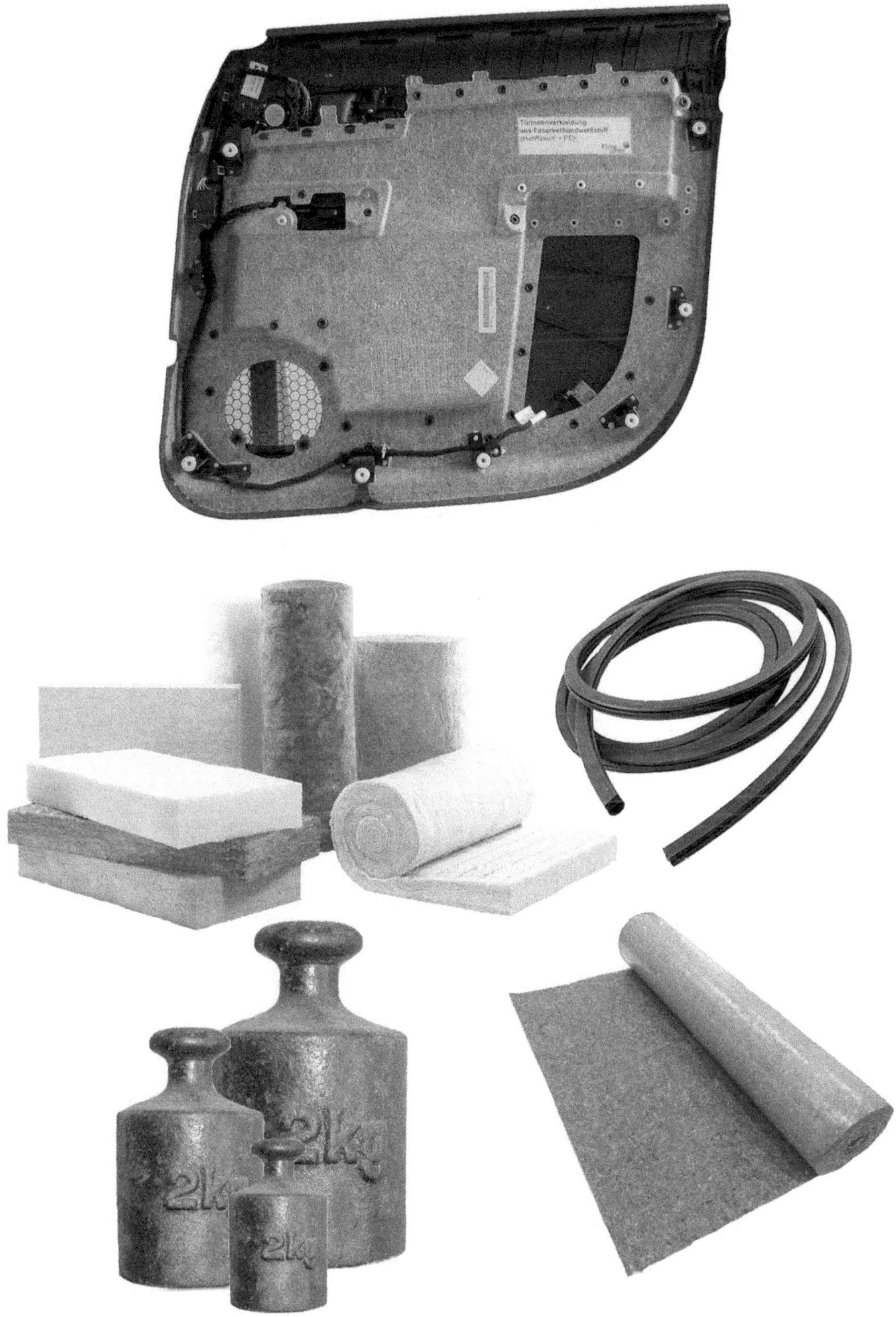

FIGURE 7.5 *Designing sound at times means stuffing material into a car door. The pricier your car the more silent it is inside, the more solid are the built-ins (Bijsterveld et al. 2014).*

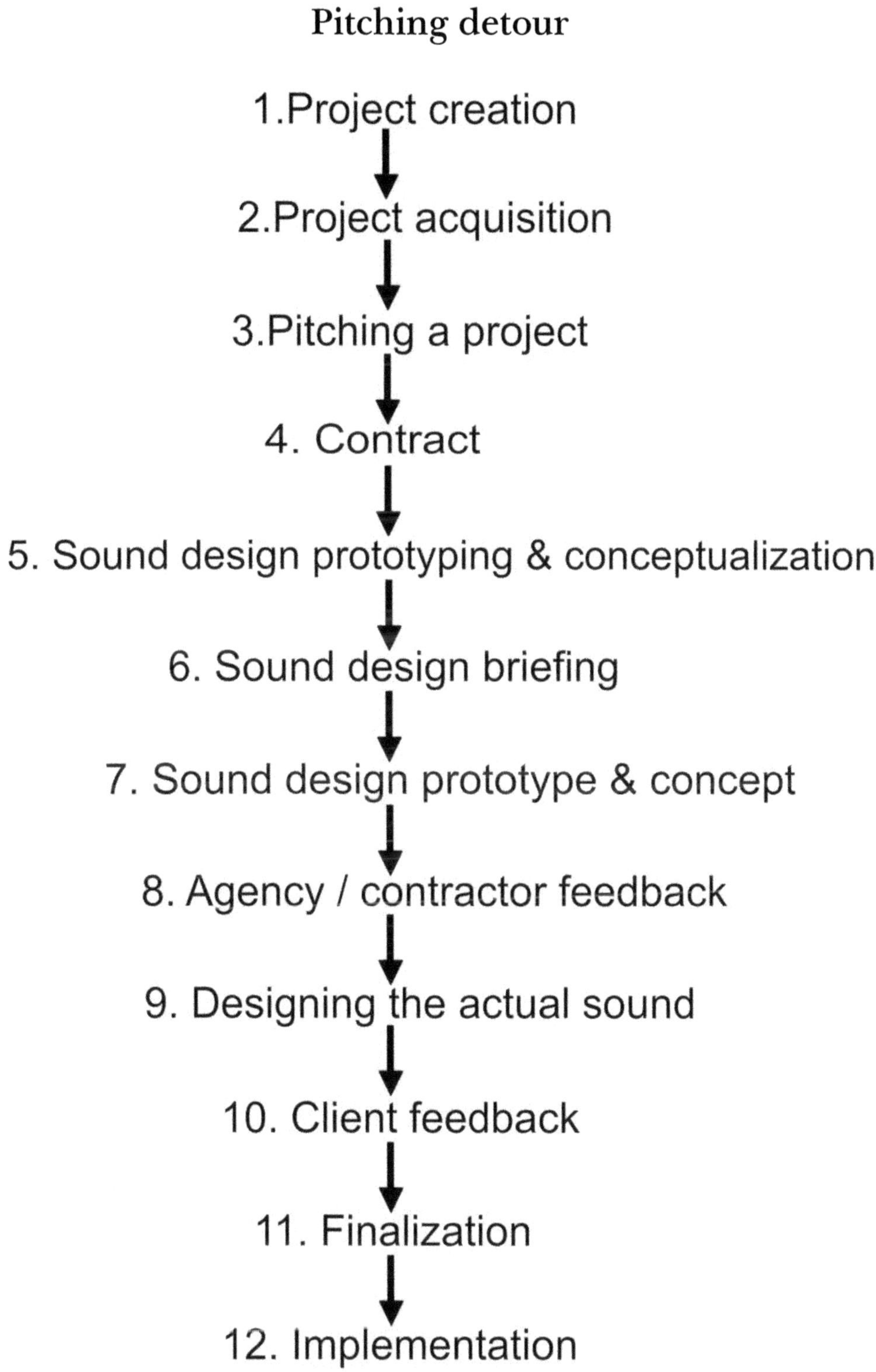

FIGURE 7.6 *How is your workflow organized? Do you recognize your actual work in this textbook flowchart? If not, create a different flowchart!*

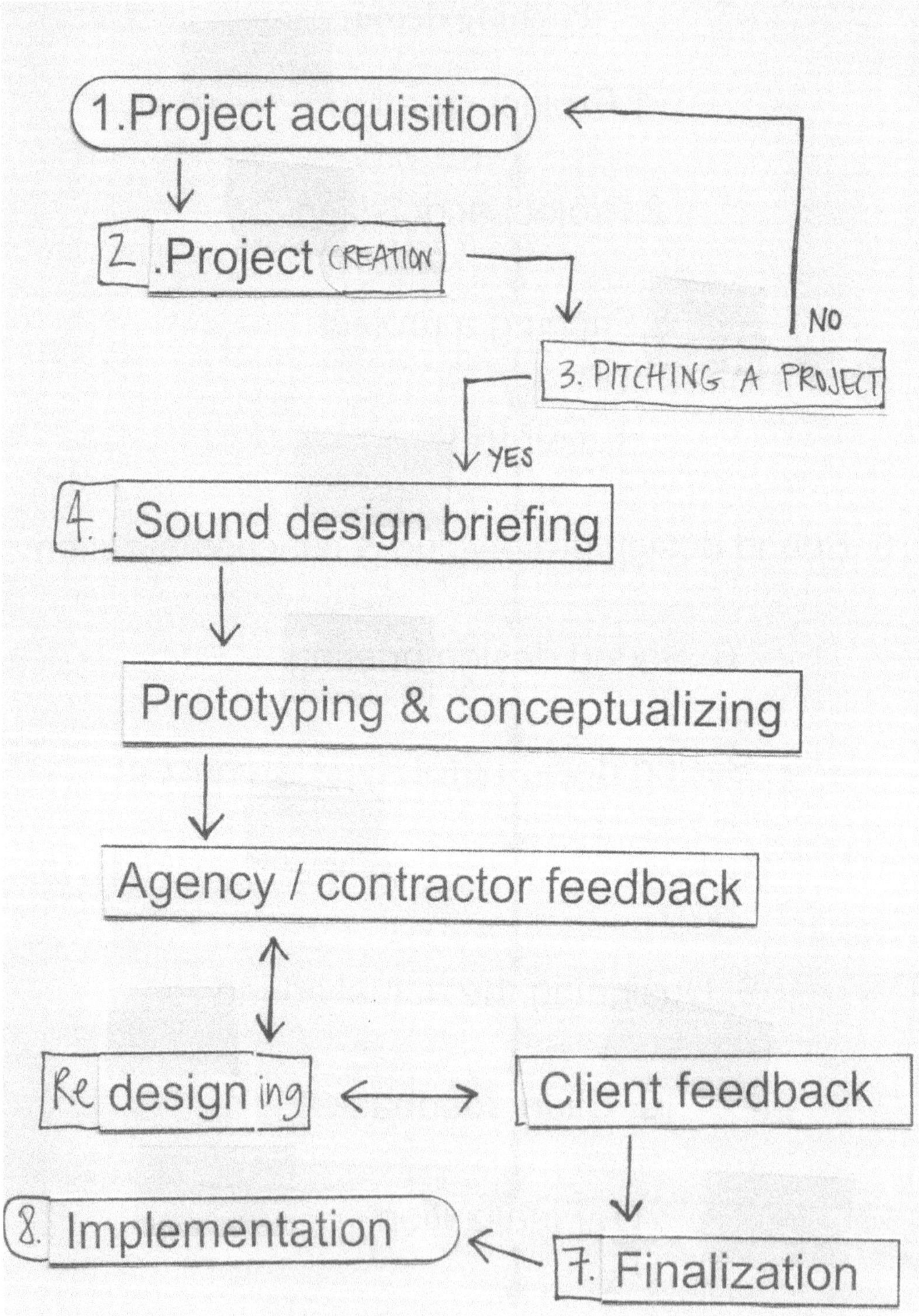

FIGURE 7.7 *A sound design workflow with some common loops and feedbacks built in: pitching and acquisition, contractor feedback and redesigning, client feedback and finalization.*

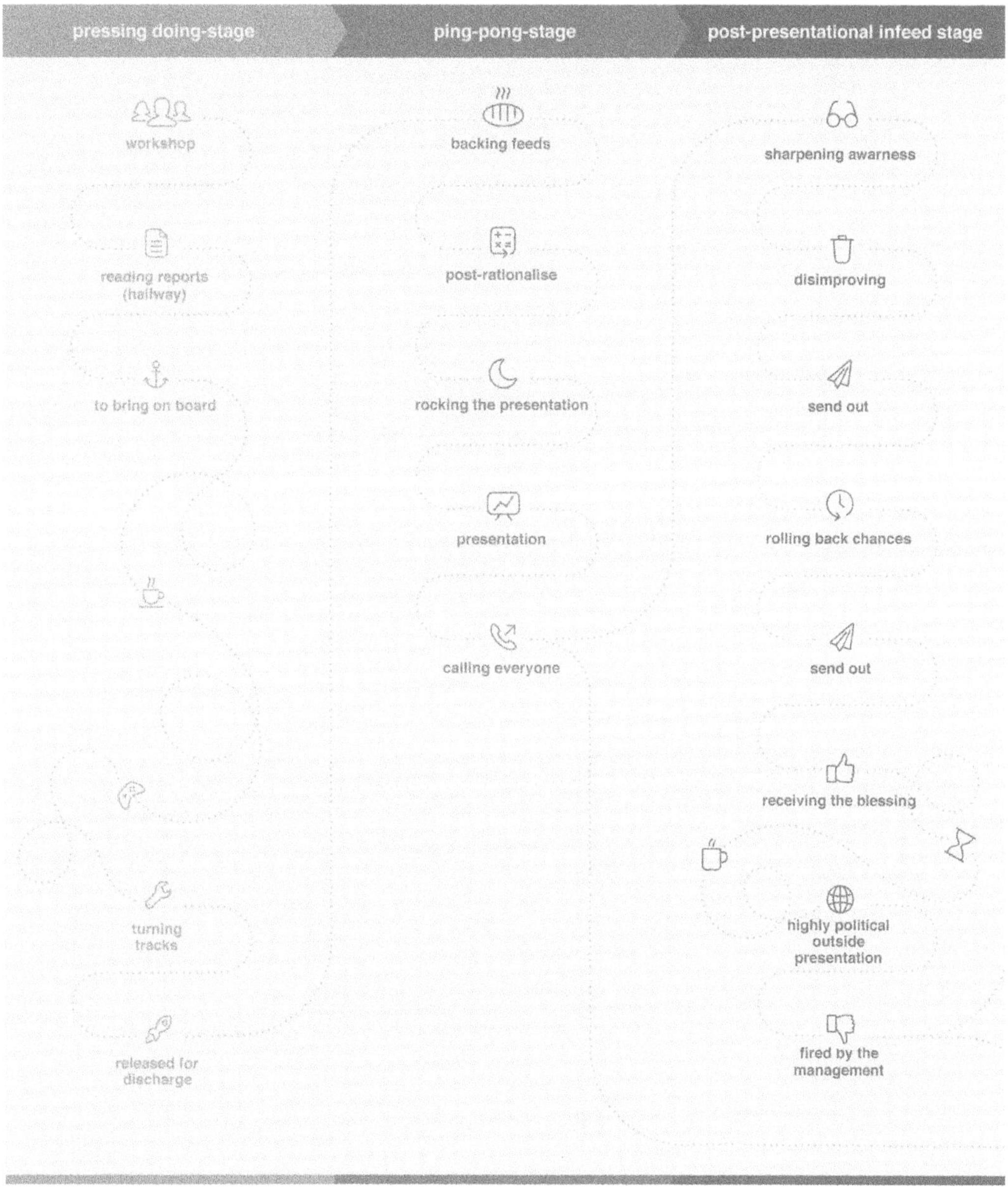

FIGURE 7.8 *One sound design agency developed this process flowchart in six stages from the "pressing doing stage" to the "postfinale stage"—with a good amount of realism, sarcasm, and self-deprecation thrown in.*

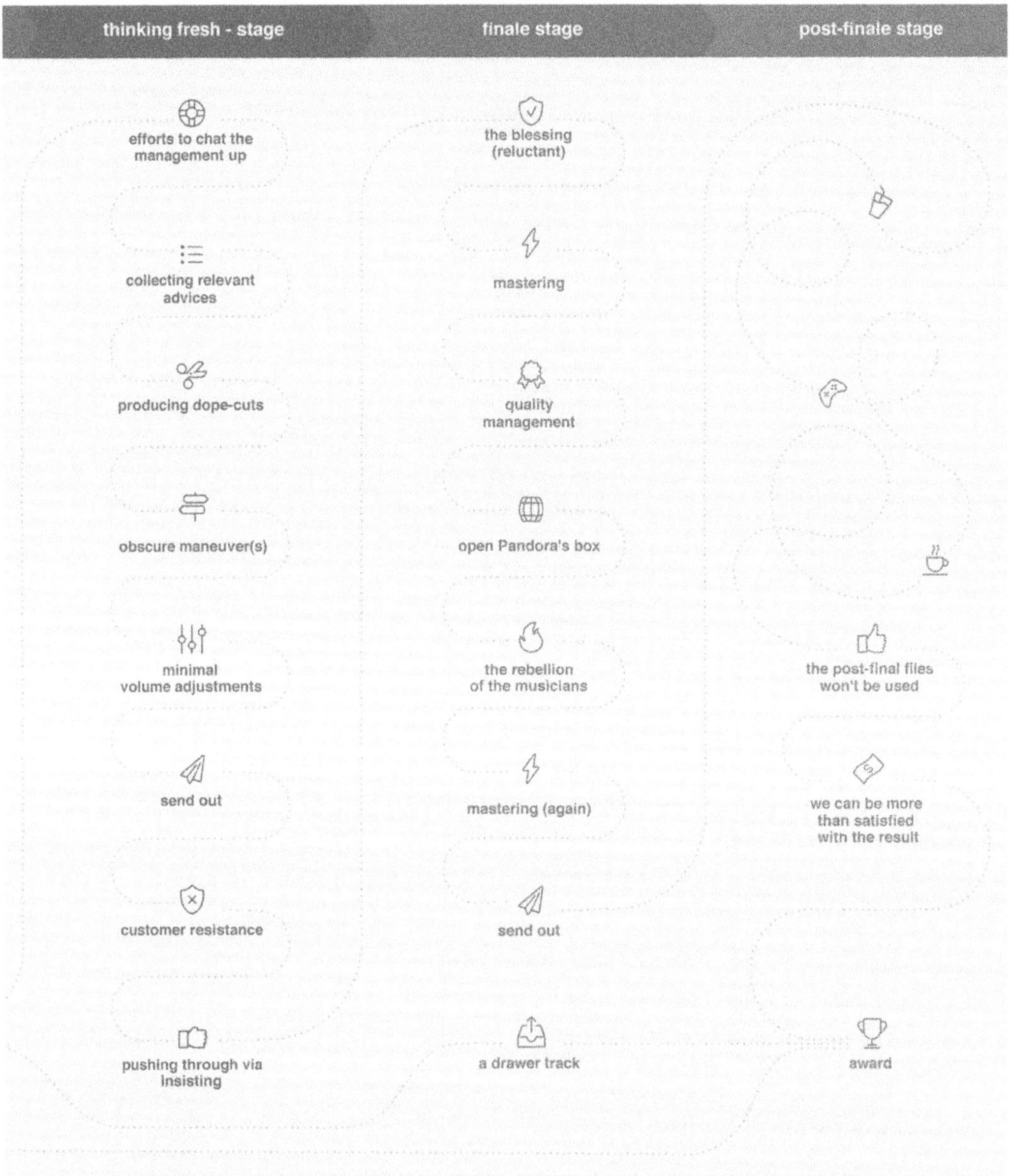

FIGURE 7.9 *Is this the final flowchart for* all *creative industries? (Picture reference: Leopold Hoepner & why do birds.)*

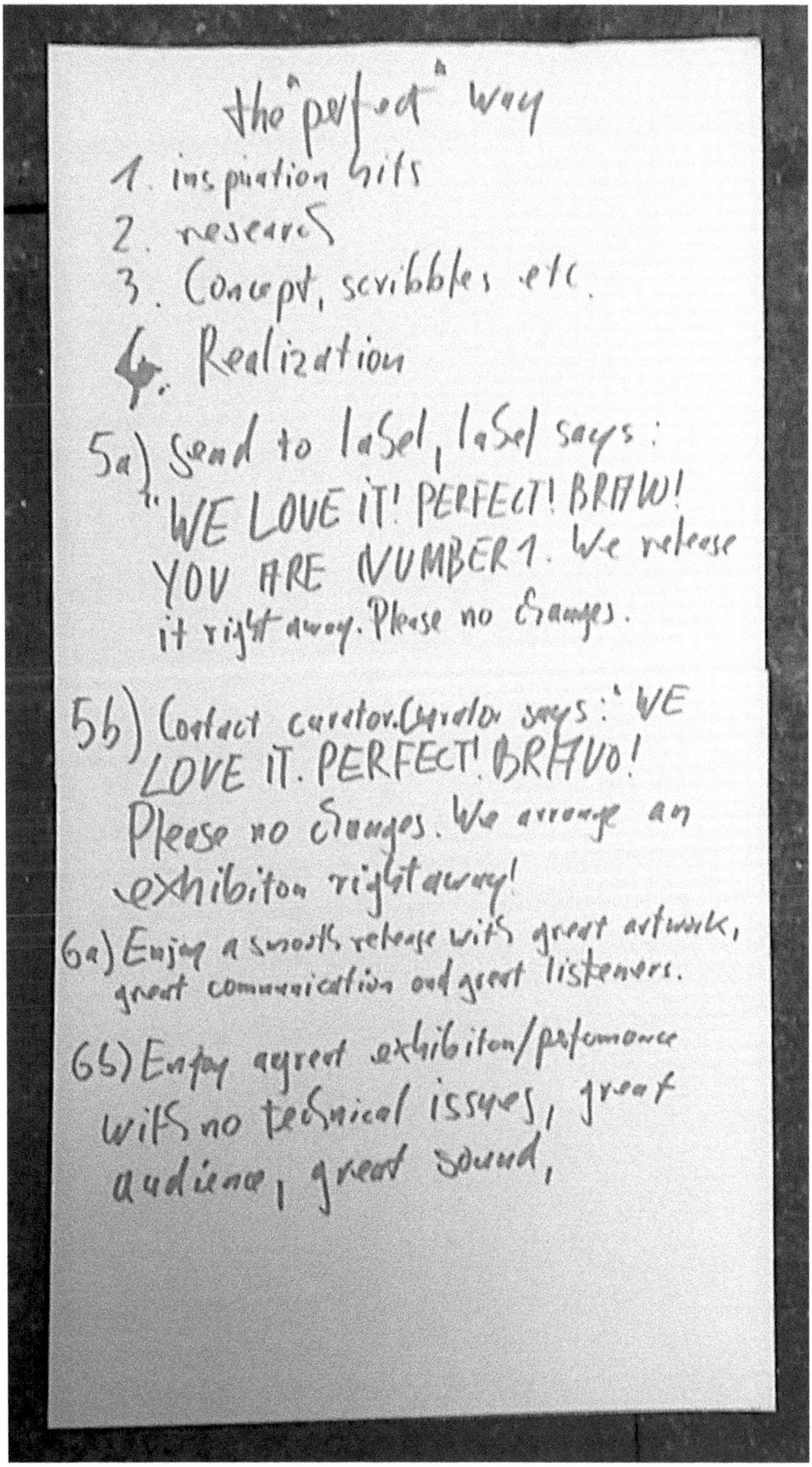

FIGURE 7.10 *The perfect way. An artistic vision, a desire for* sonic excess *as well as for public recognition. (Picture by Robert Heel.)*

Design abuse

FIGURE 7.11 *What use did you have in mind when designing? And what did the* consumer citizens *do with it anyway?*

FIGURE 7.12 *Search in the internet, look for "Pianist angry by Nokia tune," "All star but it's played on an old samsung phone," "iPhone 7 remix," or the "Cellphone Bird" ...*

FIGURE 7.13 *Who doesn't mute button tones or game sounds on their mobile device? And if you never did: how many fellow passengers felt the urge to kill you on the spot? Sound designers are arms dealers in the* Sonic Warfare *(Goodman 2009).*

FIGURE 7.14 *Sound design is an ancient craft, like this ancient pattern called "Circle of life." Instrument makers, singers, conductors, or composers are their ancestors. But with a little bit of help by* innovation *you stole from them, willfully or unaware of it, no doubt.*

FIGURE 7.15 *And sometimes it's just better to destroy something to come to a new idea or a new path of designing.*

Sonic darlings

FIGURE 7.16 *What and who are the latest sonic darlings of today? Our sonic darling is Elly Heuss-Knapp one of the various inventors of the radio jingle. Trained as a teacher at a girl's school and deeply concerned with civic education in interbellum Germany, Heuss-Knapp became a frequent public speaker on women's suffrage. In 1933, when national socialists came to power, she was no longer allowed to speak publicly. She started working in advertising and held at the time the patent for radio jingles (drawing by Friedrich Naumann).*

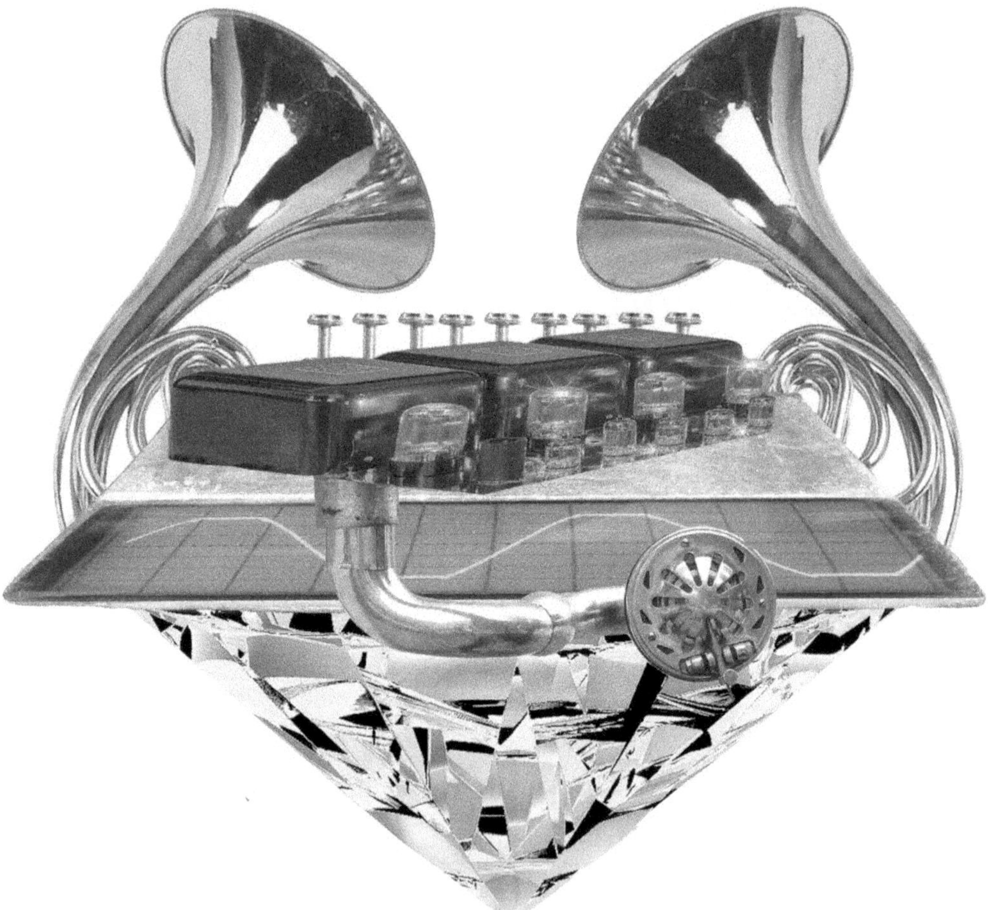

FIGURE 7.17 *No one will hear the difference, but when there is money around, we love to go HIGH END. An absolute crazy set up, a golden playground.*

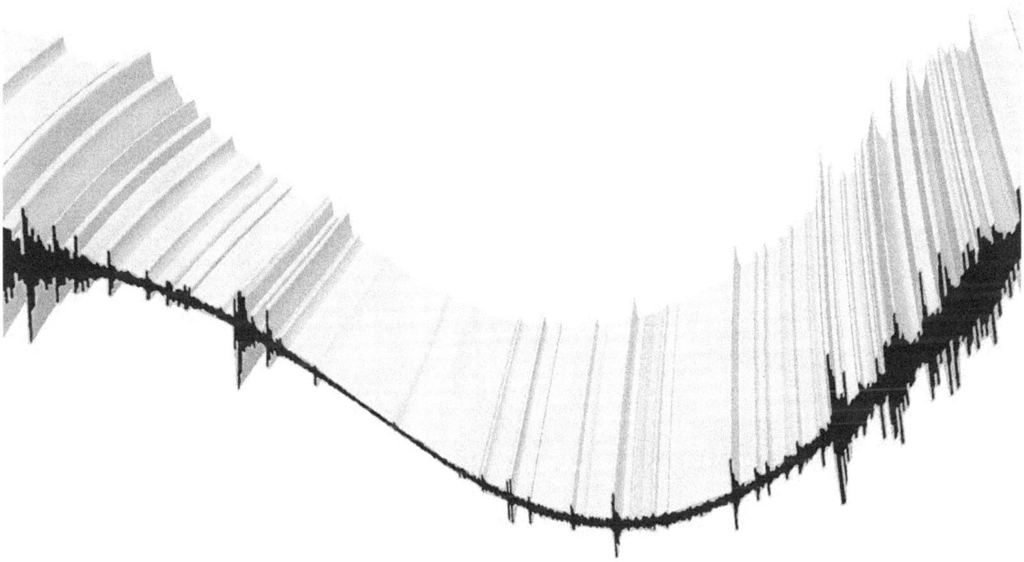

FIGURE 7.18 Wave Field Synthesis, Spatial Sound Systems, 3D Sound: *what do you actually hear when you listen to sound being projected "as in real life"?*

FIGURE 7.19 Minimal is the new maximal. *After colorful and rich ambient atmospheres for years, the new jingle is a reduced and concise logo:* Flat Sound Design *(in the 2020s, though, the pendulum will surely swing back).*

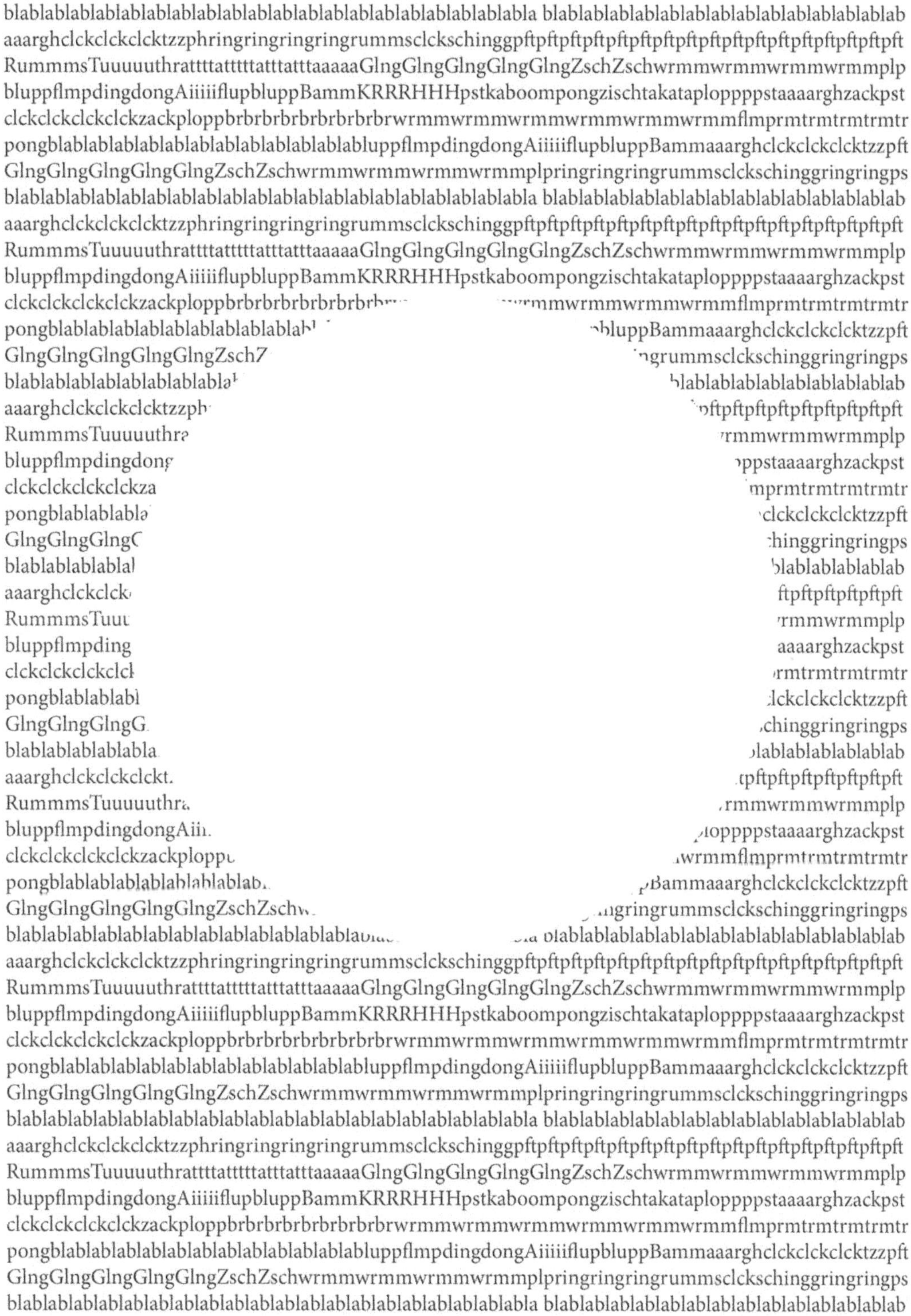

FIGURE 7.20 Silence is sexy *(Einstürzende Neubauten). A bit of nothing gives the final touch to your design.*

8

Portfolio and Presentation

Gray minimalism

FIGURE 8.1 *Sound design websites are white. Ninety-five percent of them at least. What can you do with a white template?*

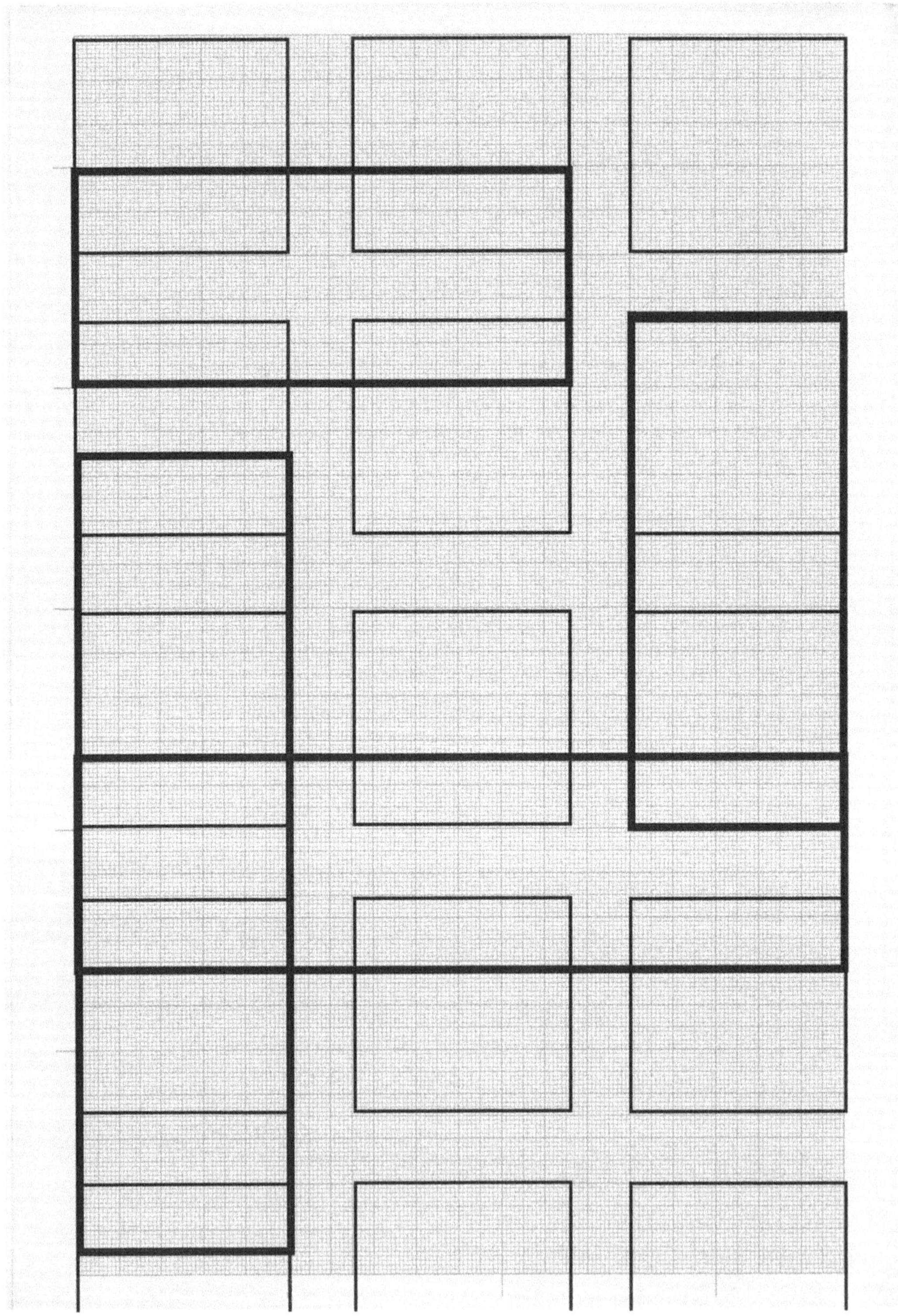

FIGURE 8.2 *Sound design websites prefer a neatly sorted, modular arrangement of brief paragraphs and larger, suggestive images.*

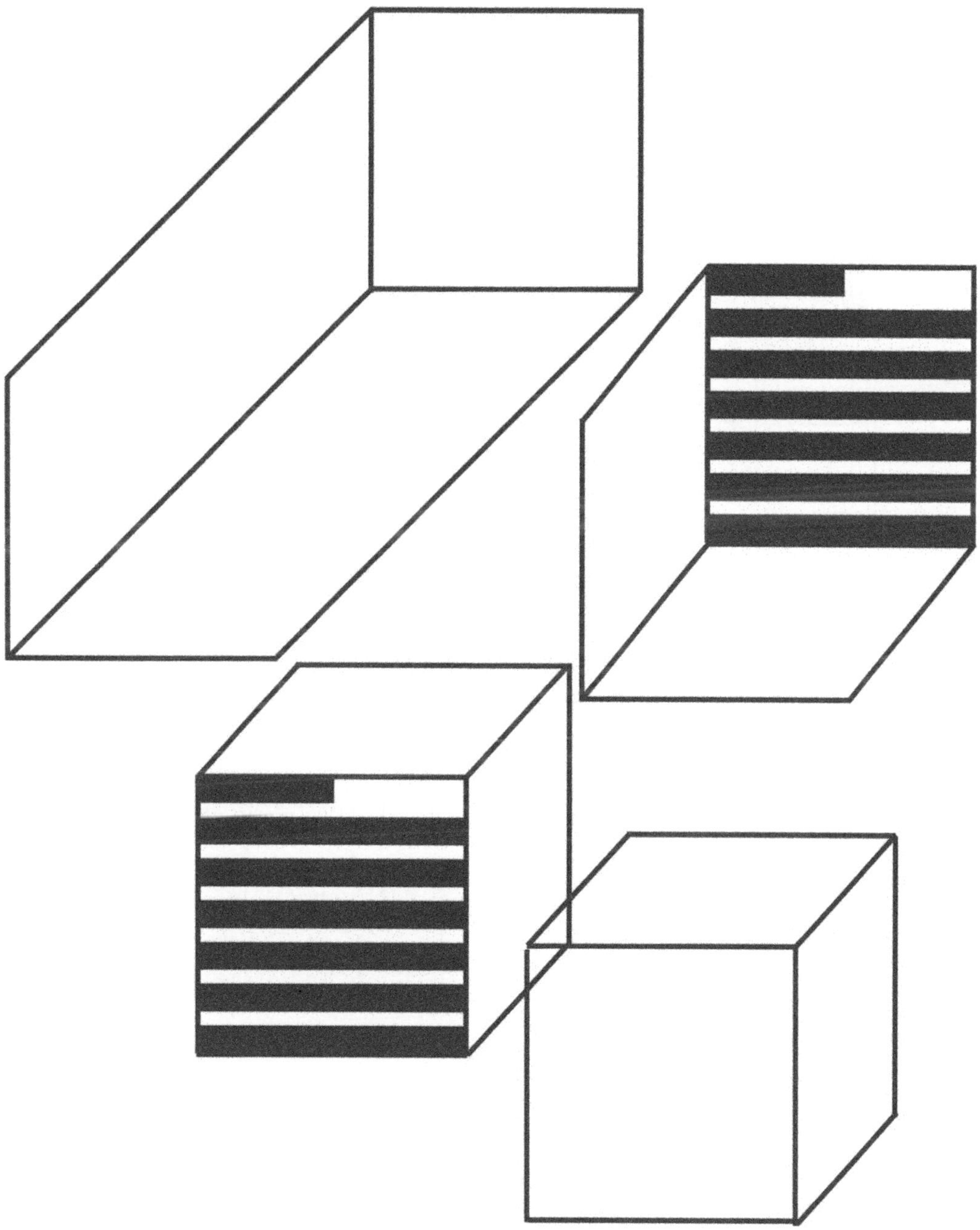

FIGURE 8.3 *Sound design websites love boxes. Flexible, easy to access—and so* Wordpress*-able!*

FIGURE 8.4 Sans Serif—or die! *The 2010s indulge in this typeface family from the 1910s: still fresh, aggressively contemporary, just slightly futuristic.*

FIGURE 8.5 *Sound design image tropes include rural animals, earnest adults (with glazing eyes), young women and kids wearing headphones, and a drop falling into water (with the oh so inspirational concentric waves).*

Staging airwaves

FIGURE 8.6 *How to represent sound design in pictograms? A loudspeaker, a micro- or headphone, an equalizer display, the right human ear, a violin clef, or two eighth notes beamed together. The visual repertoire seems limited.*

FIGURE 8.7 *Circles and discs, lines and dots represent the process of designing sound. But seriously: why?*

FIGURE 8.8 *Circles and discs, lines and dots are then connected to represent wheels and cakes, clusters of molecules, and symmetrical figurations of circles and discs. Nice, peaceful, and cute, but, still, why?*

FIGURE 8.9 *Some verbal tropes of sound design: sound is* emotional, *the* concept *represents* innovation, *the* brand *has a* value, *the design* communicates identity.

FIGURE 8.10 *Do these words still mean something?*

Selling the sweat

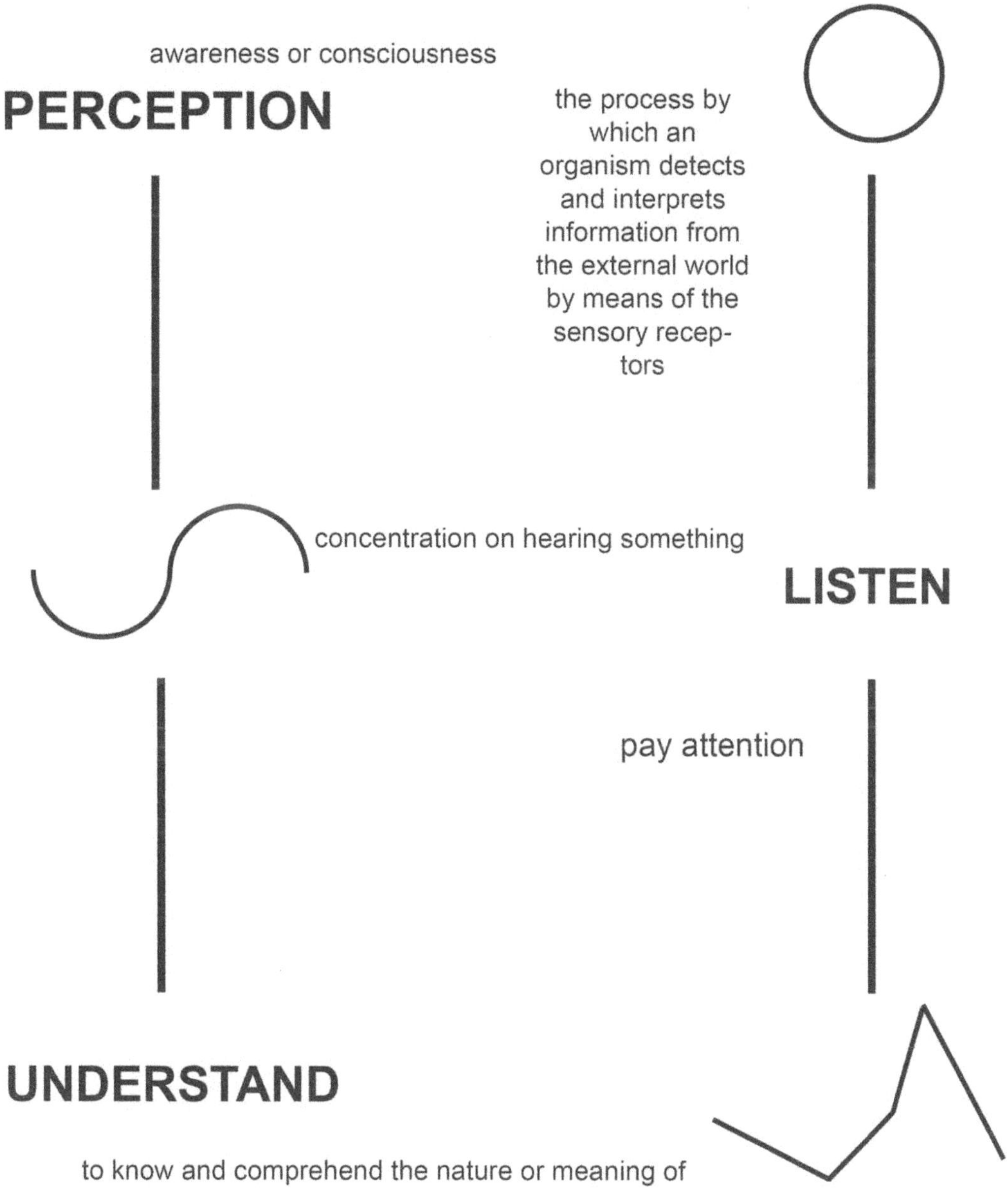

FIGURE 8.11 *You need to sell the sweat of yourself working hard. So, how many steps does it take to develop a sound design? Five, three, or twelve?* THE ANALYSIS *is always the starting point (design also needs to resemble an intellectual or even hermeneutical endeavor these days).*

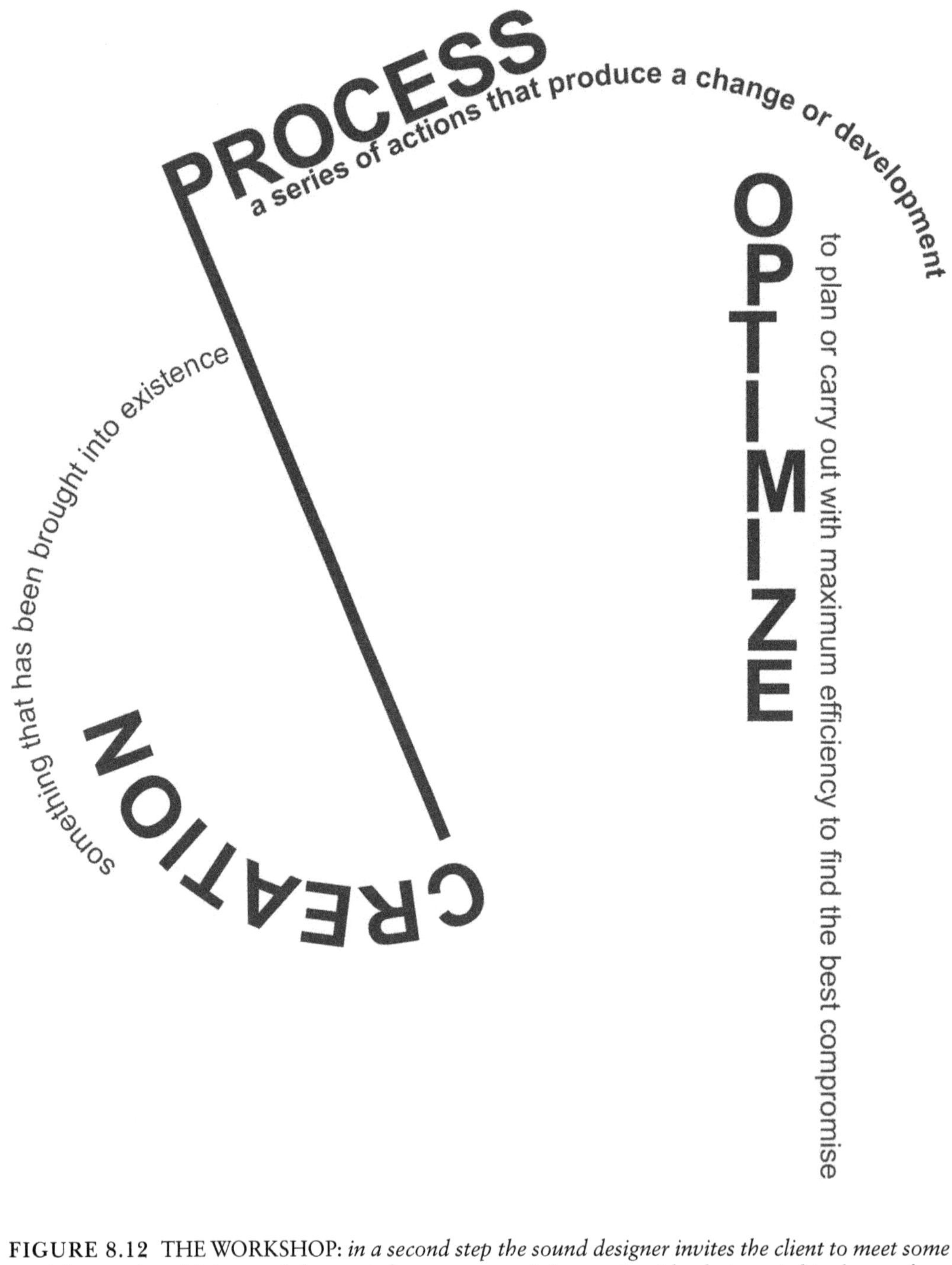

FIGURE 8.12 THE WORKSHOP: *in a second step the sound designer invites the client to meet some musicians, artists, thinkers and the magic happens: organic innovation (the designer is hired to perform this magic trick)*. Process, creation, optimize.

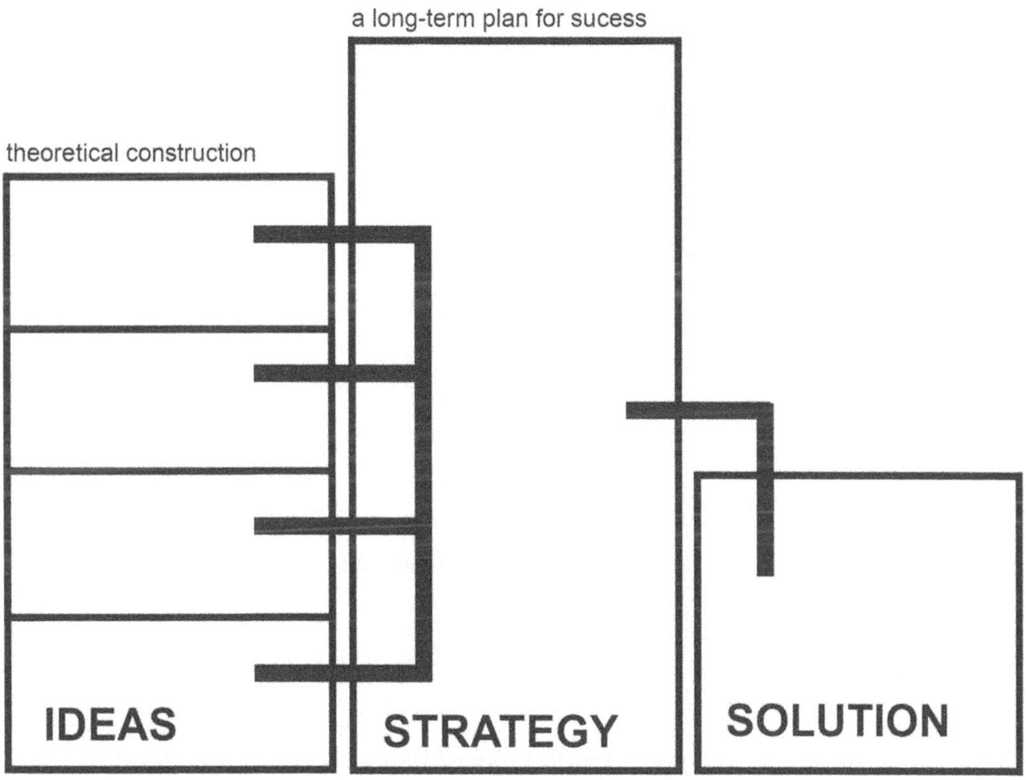

FIGURE 8.13 THE CONCEPT: *it needs big words, big blocks, big, clear, capital letters. Keep it simple, stupid.* IDEAS. STRATEGY. SOLUTION.

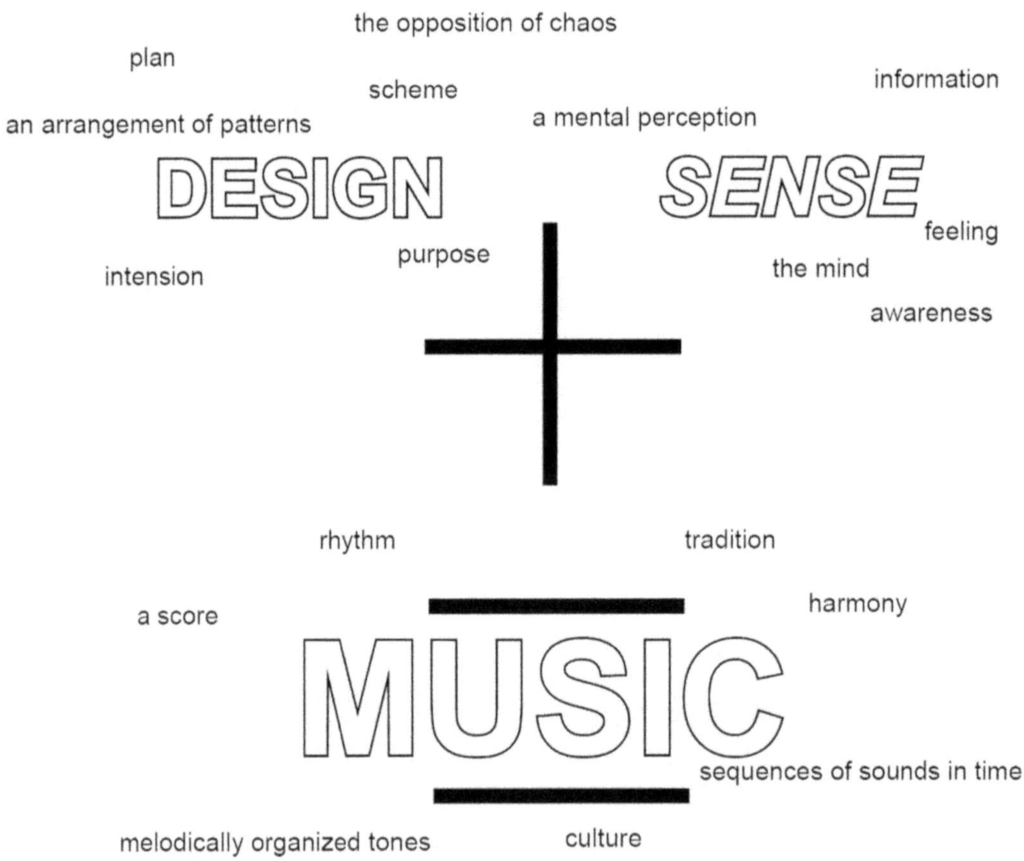

FIGURE 8.14 THE PRODUCTION: *the production team comes together:* DESIGN + SENSE = MUSIC. *The result of their work is presented as a well-organized and well-documented process (remember: the designer is hired to comfort the client).*

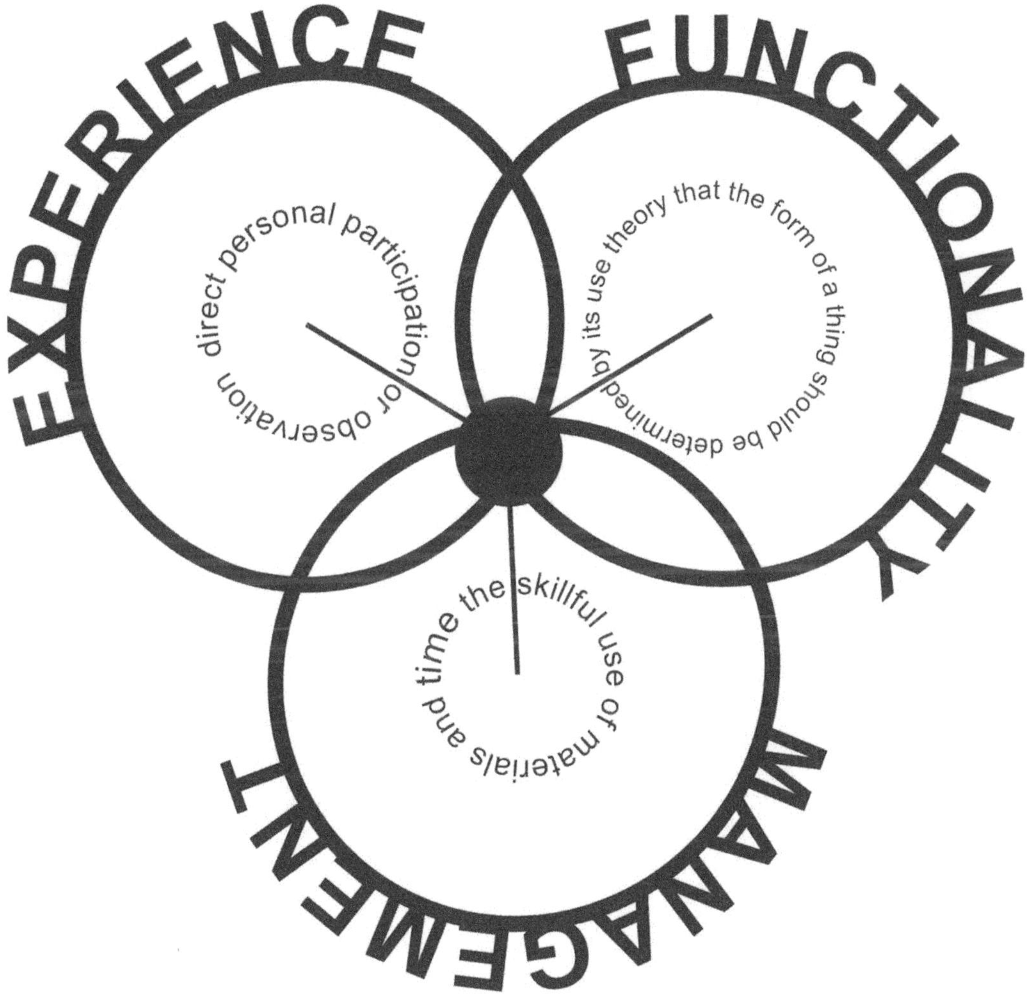

FIGURE 8.15 THE IMPLEMENTATION: *after finishing the design the hard work of implementing begins*. Your experience counts! It's so functional! You can manage it! *But does the company really care?*

Channeling the shaman

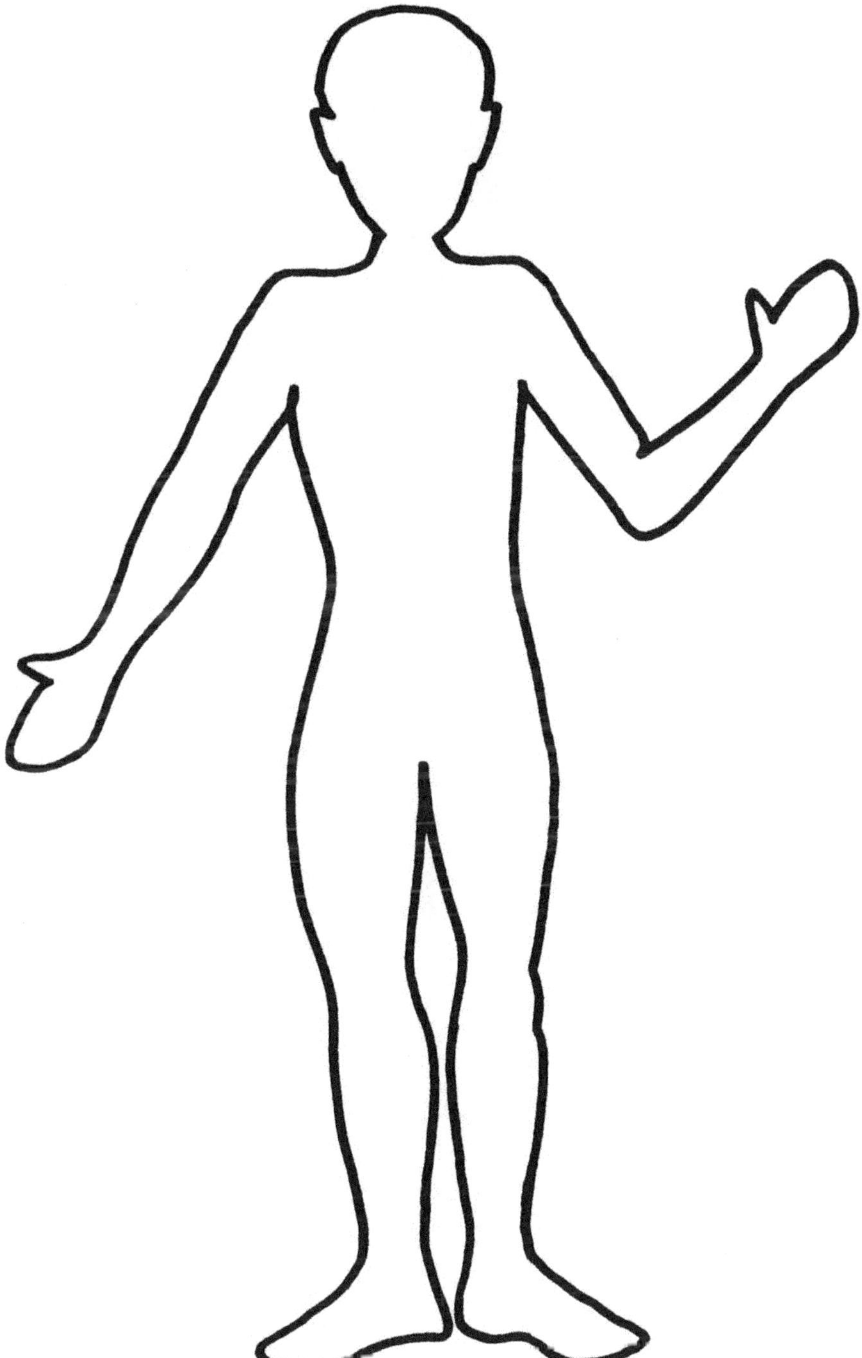

FIGURE 8.16 *What does a sound designer, after all, look like? A shaman, a funny guy, a slick consultant, or a freaky weirdo?*

FIGURE 8.17 *If you are a musician you might be in constant company with your favorite instrument. A trucker cap and a full beard are recommended in the 2010s (subject to change; gender apparently still retro-male).*

FIGURE 8.18 *If you are the manager-type the tie is mandatory, a decent haircut as well. A suit is not bad. Polish your shoes!*

FIGURE 8.19 *If you consider yourself the technician-type, you will choose a more inspirational haircut. Sunglasses for extra weirdness.*

FIGURE 8.20 *You can be* The Artist *(with a big T and a big A!), which grants you free choice out of all possible accessories, garments, and looks. However, one accessory seems mandatory (in certain phases of your career). AND WHY ON EARTH ARE ALL THESE GENERIC DESIGNERS STILL ONLY MALES?*

PART THREE

Living With Sound:

The Semiotics and Mediology of Sonic Signs

Written in collaboration with Carla J. Maier

9

Signifying Sounds

The semiotics of functional sounds

A sonic pattern language?

Walking along Zossener street in Berlin, I hear the rhythm of drums coming from the front door of a house on the opposite side. A group of women, men, and children dressed up in festive garments come out onto the street, accompanied by a group of musicians. Sounds from a drum—a *davul*—a tambourine, and a flute—a *zurna*—can now be clearly associated with Turkish wedding music. The music, together with the singing accompaniment of the band members and the hand clapping and singing along of the wedding crowd, fills the street and creates an energized, euphoric, and festive atmosphere. Along the sidewalk, a row of fancy cars is lined up, one of which is elaborately decorated so it can be identified as the car for the wedding couple. The main attention is drawn to the cars on the left and right of the wedding car, in which young male wedding guests activate the car horns and rev the engines of their cars in celebration of the happy event. The sounds of the davul and zurna are still audible, but only temporarily as a rhythmic and atmospheric noise in the background, while the car horn sounds press to the foreground. The clapping of the wedding guests competes with the car horn sounds, which now occur in faster sequences, and the voices of the singing crowd are momentarily swallowed up by the noise of the car engines. In the meanwhile, the wedding ceremony proceeds to the wedding car, in which the wedding couple finally take their seats. The other wedding guests then proceed to their cars, and while the musicians are the last remaining on the sidewalk continuing their songs, the car parade begins to move. Finally, the musicians also retreat into a car, and the whole procession leaves the scene. It is encompassed by the sounds of car horns welling up, until they turn the next street corner and leave a whirring atmosphere behind that is slowly overlaid with the everyday noises of passing cars, the wind blowing through trees in the front yards and the sounds of a suitcase on asphalt.

The implications and meanings of the sonic signals that are part of the above description are generated within complex temporally, spatially, and culturally specific situations. The sounds of the musical instruments and the noises of the car horns are only ostensibly invested with clear and unambiguous messages: reduced to its mere functionality as semiotic signals, the music represents traditional Turkish culture, and the car horns constitute mere auditory traffic signals. Interpreted in this way, the sounds of the davul, tambourine, and zurna can be identified as belonging to *sonic patterns* that form the main musical elements of this wedding celebration. Read in this fashion, these patterns seem quite stable, as they refer to more *conventional musical sign systems*. In a similar manner, the basic functionality of the car horn is its attribute as a traffic signal to attract the attention of other traffic participants (drivers, cyclists, pedestrians), and the sound of the car engine signals the proper functioning of a car. However, an investigation into the specific *situatedness, materiality, and performativity* of these sonic signals reveals that they are highly ambiguous and resist any form of simple categorization in terms of *conventional codes* and clear messages—at least according to the stricter definitions of semiotics.

It is in this ambiguous and conflicting sphere that people have to negotiate sound's rigid functionality and the affective, spontaneous, compelling, or disconcerting effects of sonic events in everyday life. *Living with sounds* occurs exactly in such blurred and often paradoxical situations. The sounds depicted in the above description generate a multilayered sonic event, an evanescent and ambiguous situation, an individual and collective experience, and an act of transcultural transmission. This chapter hence studies this wedding celebration as a result of highly ambiguous sign processes and erratic situations, communications, and thoughts, while taking into account, extending, and revising theories of sonic signification.

In order to enter into the realm of sonic perception and interpretation in relation to sonic signals in everyday life and their implications for sound works, Christopher Alexander's theory of *pattern language* (Alexander 1977) offers a useful analytical concept to think about *sonic patterns* and a *sonic pattern language*: sonic patterns are defined for our purpose here as a way of interpreting physical and perceptible sonic signifiers and vernacular sound practices, and thus a sonic pattern language would then refer to a comprehensive method of understanding and interpreting functional sounds, and of approaching sound design problems from this interpretation: examples of arcade sounds, of elevator music, or of a lightheartedly named "Joe Voice" (see Figures 6.1–6.10). This transfer is close to Alexander's understanding of design patterns, as explicated earlier in this book (in Chapter 1, section "Dialectics of functionality"). Thus, the aim of this chapter here is to further elaborate the analytical work on particular sonic situations and amplify it toward a more *general and cultural sound design theory*.

In his book, Alexander highlights—through his very detailed ethnographic description—how moments of social interaction on occasion of an entrance hall of a house, for instance, are spatially organized: how certain architectural patterns significantly influence the ways in which people act and interact. This deeply anthropological approach to design is applied in this chapter to the example of the urban wedding and extended through an analysis of the recurring *sonic* patterns that—similarly to Alexander's architectural patterns—resemble and evoke social activities. One might begin with functional sounds emanating from car horns and car engines that belong to the sonic pattern of *traffic signals*, go on to the musical sounds that accompany the wedding ceremony that belong to the sonic pattern of *wedding music*, and the cheers, laughs, and claps of the wedding crowd that can be summarized as a sonic pattern of *celebratory mood sounds*. In an even more extended understanding of a sonic pattern language either emphatic reactions by passersby, such as cheers, claps, and laughs, or the more disregarding exclamations of annoyance, or mere muteness, can be ascribed to a pattern of *participatory listening*—an agile listening that complements the meaning production of a sonic situation. Other *background sounds*, such as wheels or footsteps on asphalt, the wind that blows through trees, doors that open and shut, form another sonic pattern. The sonic events collected here as part of an ethnographic description of a sonic situation resemble a cluster of sonic patterns, for which architectural patterns such as "pedestrian street," "car connection," and "hierarchy of open space" (*cf.* Alexander 1977) set the stage. Pushing this multidimensional and multisensorial perspective on sonic patterns further, Alexander's architectural design patterns work as complementary elements to the sonic patterns, thus emphasizing the spatial dimensions of sound design and the spatiality of sonic patterns. Consequently, the resulting *sonic pattern language* that consists of consciously designed sounds and conventionally acquired listening practices, is equipped with its own affordances. In reference to William Gaver's notion of *technological affordances* (as introduced earlier, in Chapter 1, section "Dialectics of functionality"), which help to analyze "factors of perception and action that make interfaces easy to learn and use" (Gaver 1991), *sonic affordances* extend the scope of the interface to include an assemblage of sounds, technologies, perceptions, and actions that could aptly be described with Rolf Großmann as an *auditory dispositive* (Großmann 2008; Schulze 2016b). In this regard, the sonic patterns described above—traffic signals, wedding music, celebratory mood sounds, participatory listening, background sounds—signify in relation to each other: the car horn sounds afford celebratory, joyful, and boisterous moods when set against the background of a wedding ceremony. The wedding music affords the "hierarchy of open space" to shift by gaining sonic dominance (Henriques 2011): the intensity of live music, bodies in motion, and the focus on the newly wedded couple creates a kind of enclosed, almost private space within the public space that at the same time remains porous, as passersby get infected by the music and join in the celebratory mood sounds. The background sounds that are present at this point in time on this particular pedestrian street afford a sonic stage for the wedding ceremony to take place, a stage—when left by the wedding crowd—that remains filled for a while with the whirring atmosphere that is a remanence of past sounds, materials, movements, and actions. In this way, a *sonic pattern language*—as part of more encompassing sonic sign systems—signifies beyond a structuralist focus on language, or text, as such: it allows a focus on the perceptions, usages, and interpretations of functional sounds (as sonic signs) in specific social situations. In this way, functional sounds regain humanoid and situational agency, they rework social and also cultural hierarchies and they thus resist and transform formalized and rigid functionality. Now, how does the concept of sonic pattern language work not only as a methodology to interpret functional sounds in everyday life, but also as a tool to approach sound design problems? For instance, as our ethnographic study of working processes in sound design has revealed, there are a number of entangled and sometimes also dissonant trajectories that result in the creation of sound: the language that is used to describe sounds (Figures 6.6–6.10 and Figures 8.6–8.10), the tools and devices that are used to generate sounds (Figures 5.6–5.10), and the working environment within which sound is created (Figures 5.1–5.5), as well as the forms of how the sound design process is staged and mediated (Figures 5.2–5.5, 5.16–5.20 and Figures 8.11–8.15)—all these elements constantly oscillate between conventionalized codes, individual life-worlds, aesthetic preferences, and social hierarchies. Although functional sounds are *constituted* by these complex assemblages, this complexity is hardly ever reflected in any straightforward way in the actual functional sounds we hear. This can be exemplified, for instance, by how a "Joe Voice" (Figure 6.7) became part of the verbal repertoire to describe a specific style of the voice that the sound designer should apply to a character of a children's game app. In this context, the term "Joe Voice" refers to a whole assemblage of musical socialization, insider jokes, and the knowledge of sonic qualities and how they might be pushed so as to suit the character of the jolly construction worker who is part of the children's game sound. Referring back to the discussion of sonic patterns, if a sonic pattern language included sonic patterns such as "Joe Voice," along with more common patterns such as "arcade sound" (which refer to 8-bit sounds) or "elevator music," this terminology might enrich sonic thinking and communication in sound design.

Rematerializing sonic signs

In our analysis of the previously narrated wedding ceremony, sonic patterns are interpreted as recurring and conventional signifiers that are part of culturally specific everyday practices. In concurrence with Alexander's notion of design patterns, *sonic patterns* are evocative and suggestive rather than prescriptive and commanding, and thus leave space for the unfolding of ambiguities and, as will be elaborated later in this chapter, for transcultural negotiation and aurality. Sonic patterns such as *mood signals, wedding music* or *traffic signals* are open categories that are not necessarily constricted to "Western" musical or functional sonic traditions or conventions. The sonic patterns come to life—and become rematerialized—only when applied to a specific sonic situation.

But how can sounds be interpreted as ambiguous, arbitrary, and unstable signifiers, how do sounds become part of an activity? Thinking beyond the structuralist perspective on signs as promoted by Ferdinand de Saussure (to which after all even traditional cultural studies approaches refer) and rather than following the ways in which his approach was continued in the works of Valentin Nikolaevich Voloshinov, Jacques Derrida, Roland Barthes, or Jay David Bolter (Chandler 2007: 51–55), we aim at elaborating a more dynamic, material, and situative approach to semiotics in reference to Charles Sanders Peirce's initial work of putting forth a *rematerialization of the sign*: rather than sticking to the structuralist notion that a sign has an intrinsic and defined meaning, Peirce proposes that semiotic processes consist of a *thoroughly dynamic threefold relationship* between a sign, an object, and an interpretant:

> The Dynamical Interpretant is whatever interpretation any mind actually makes of a sign. This Interpretant derives its character from ... the category of Action. (Peirce 1998: 499)

The approach to sonic signification that we develop in this book seeks to push this *category of action* even further than Peirce probably intended, extending it to the corporeal, transactional, and social dimensions of sonic signification. In this way, functional sounds become part of personal and individual as well as social and collective activity. Actions are, hence, not only restricted to the obvious physical activities that might be the result of a sign process, but also self-reflexive decision-making or decision planning processes as they occur in everyday life in those heavily networked, transnational, and globalizing societies of the early twenty-first century. The truly generative aspect of Peircian semiotics for our issue here is that there is not just a simple and dyadic relationship between *"The Sign"* and *"The Object"*—strangely and erratically distinct from each other—from which an unambiguous message is derived. Such an atomistic and dyadic relationship would not be sufficient for describing sign relations in a sinuous world with all its allusions, undertones, hidden agendas, forgotten but still sensible interpretations as well as highly relational aspects tying a sign and its interpretation to a person living with this sign. Hence, according to Peirce this relation is malleable and can change and transform rather easily. If we transfer Peirce's expansion from the area of the verbal to the area of the sonic we must state: *depending on the specific situation in which a sonic event occurs, the sound of this event signifies a meaning in relation to this situation; thus, this sonic sign can and must also be interpreted as a genuine and intrinsic part of this specific situation.* Any *sonic signifying process* is therefore to be investigated as a process in a particular auditory situation in which humanoid aliens like you or me refer to, draw upon, and produce particular sonic events. Materiality and situativity constitute in confluence the meaning at play here.

The Peircian notion of the materiality and situativity of any signification process provides in consequence a whole new perspective of how to interpret sonic events that interrupt and accentuate our everyday experiences. And it also might, furthermore, add a crucial insight to the various major efforts to constitute a so-called *Semiotics of Music*, a *Semiotics of Sounds*, or a *Theory of Musical Meanings* or *Sonic Meanings* (e.g., Nattiez 1976, 1987; Tagg 2012): the signification of sounds is generated within a situated context and they receive their meanings out of a specific interpretation of the sonic materials in relation to an individual listening, sounding, and performing biography of the listener. Take the various sonic agents at the wedding ceremony discussed at the very beginning of this chapter: the meanings of the functional sound of the car horn or the instruments; the dancers and their cheers toward the sonic patterns, and how they shift accordingly; last but not least my very own ethnographic perspective being part of the wedding ceremony. All these individual hearing perspectives are all the time also a corporeal, communicative, medial, and mediated activity. The *interpretation of sounds* therefore needs to be rematerialized. And this is true for the realm of music as well as nonverbal, functional, and vernacular sounds. A theory of musical and sonic signifying processes, and semiotics in general, is therefore—as already tentatively stated by some of the researchers in this field—merely possible as a *performative project*, not as a totalizing project of collecting and registering all the possible meanings and their connections to certain sounds. Such a totalizing project would surely have been a desired if not craved one for a number of nineteenth-century scholars in musicology, and this urge can even be found in the contemporary field of commercial sound design for brand sounds, sound logos, and sonic environments for stores or other commercial purposes. As Brandon LaBelle analyzed it in *Acoustic Territories* (LaBelle 2010) with regard to the sounds in a shopping mall, the sounds in such a highly refined edifice constitute a specific corporeal experience that induces and incites sense-making in regard to the consumption of commodities and brands:

> The mall forms a complex audible perspective. The perfuming of its corridors with sonic matter, in conjunction with the ambient din and social exchanges, give way to a politics of acoustic space, where decibels often exceed local ordinances, and the labor of listening is balanced between social integration and monetary expenditure. The mall then is an amalgamation of multiple economies, embodying an acoustical tension pitting vocal fatigue against background music, auditory advertising against the joys of listening, and the technologies of production against the technologies of consumption. (LaBelle 2010: 186)

While nineteenth-century music scholars tended to construct the composer as genius whilst the listener was to surrender to his spell, in contemporary sound design, the designer seems equally equipped with genius power to induce the sonic consumer to feel affiliated and identify with a brand, a product, and a "whole way of life" that the catchy sound logo or the exciting background music in a clothing store make believe. But whereas the consumer appears in these theories often as not much more than an empty shell, ready to be filled with sonically induced meaning (e.g., Treasure 2010; Groves 2011), it is more realistic, and way more compelling, to describe the sign processes at play here as emerging out of highly mixed and ambivalent, contaminated and relational processes. This becomes very obvious in moments of intentional irritation of a sonic regime of consumption, for instance, through interventions such as a flash mob. I am thinking here, for example, of a Christmas carol flashmob in a shopping mall or a street dance flashmob featuring McHammer's *Can't Touch This* in a clothing store in LA. Both performances let certain meanings and areas clash—and hence generate new meanings (which is why such techniques are also easily applied by brand agencies). A performative interpretation of signifying, hence, rematerializes the sonic sign and grants it a *capacity of agency*. Almost all aspects—and this would be our major contribution to this question—of any listening situation can play a major role in understanding and reacting to sound as signifying in a given context. This might sound awkwardly simplifying; yet, actually, it is indeed awkwardly complexifying and multiplying the task of interpreting a meaning of sounds or music.

With regard to the aforementioned wedding ceremony, already the sheer sonic materialities—the celebratory enhancement of the car horn sound that is generated through penetrant repetition, and the car engine's hauling bass frequencies—detach these sounds from their conventionalized functionality as traffic signals. Functional sounds have been rematerialized here and this materiality situates the sound in a different imaginative framework. In this rematerialized form, a sign then becomes part of the continuous "work of the imagination" that Arjun Appadurai emphasizes with regard to how the ways in which we communicate, share, and reappropriate cultural artifacts gain significance as an *aspirational practice* (Appadurai [1986] 1996: 9). It is this work of the imagination that evokes shifts, turns, and recontextualized meanings of functional sounds and how they signify in specific situations. The car horns then become meaningful in a new way as they are used to demonstrate the presence of the wedding guests arriving in their fancy cars; together with the howling sound of the car engines, they also become part of a social and highly gendered signifying practice. The sounds of horns and the engines mean something inherently different here—at least for all the people joining this wedding celebration (while passersby might be mildly annoyed by these loud car sounds). Honking the horn in very specific continuous and rhythmic intervals, this sound signifies a certain social status (belonging to the community of wedding guests, being the owner of a car) and somewhat also a boldly staged masculinity. Are these horns not even *hyper-mediated* within the described situation of the wedding ceremony? The car horns signify *as* physicality of frequencies, *as* physical presence and dominance, *as* mediated and hyper-signalized masculinity. These sounds' materialities become entangled with all the other sounds that are present at a wedding celebration: the cheers and singing of the wedding crowd, the music played by the wedding band. The sounds start to compete with each other for attention. A conflicting relationship of different moods, interests, and actions is the result—and still the wedding situation stays intact, though now embedded in a larger social and transcultural context. Do I wish to join this celebration though I maybe do not feel part of it? Though all these celebratory practices seem strange to me? But maybe—in this urban environment—*I do indeed join in,* if just for a few minutes of listening, swinging, laughing. Then moving on to the next subway station, on with my errands.

The semiotics of hunting signals

On a weekend trip with my sister and my two young children, we are enjoying the completely calming atmosphere of a cottage outside of Berlin. While my sister is preparing dinner and my son is immersed in his play, the timer signal of my sister's mobile phone goes off playing a bugle tune that I associate with military signal music, something like the *Reveille*, an actual wake-up call. My sister and I are laughing out loud for its celebratory boldness, and my son jumps up from the floor asking: "What is this?"—"It's the alarm saying that the eggs are good," my sister replies, still laughing, and starting to imitate a horn player using her hands on an imaginary trumpet and mock-walking staccato-like through the big kitchen. The eggs are forgotten for the next minute or so, and instead we keep the melody loop running another round while my son starts dancing to it and we join in, dancing and singing along the bugle tune, loud and cheerful.

This description demonstrates another instance of the complex interplay of shifting, diverging, conflicting, and reconciling sonic signification. Once more, it brings to life the semantic shift that functional sounds sometimes undergo if they are decontextualized from their originally intended function. The ordering function of the sonic alarm is overruled by the sheer musicality and uplifting energy the tune unfolds when intervening in a rather relaxed situation open to new sounds. Thinking of a different situation in which people are caught in time pressure, workload, or social restraint, an alarm signal, as cheerful as it may sound, might result in a completely different emotional reaction or course of action. Therefore, there is a *perceived and situated quality of the sonic sign* that is not identical with the effect a sound designer may have had in mind when creating a sound for a product, nor with other context-specific implementations and uses. The agency of the sonic sign can result in annoyance or excitement, in suffering or joy, aspects we discussed earlier, in previous chapters of this book—and that the above example very precisely demonstrates.

Mobile ringtones and signals seem particularly characteristic in this regard, as they combine cold ordering power with all the affective facets of human communication and their ubiquity within the social sphere (Gopinath 2013). Moreover, what is striking with regard to the specific tune around which the above-described situation unfolds, is its direct link to *signaling in hunting and military music*, an aspect that I will elaborate in the following. Despite the long history of signaling in hunting and military music (e.g. Monelle 2006) that provided certain categories of sounds and their use, there is not much focus on these sounds or even more contemporary functional sounds in the long history of semiotics. Yet, in 1986 Czech musicologist Vladimír Karbusický proposed a number of categories to describe functional sounds—in his case, *hunting signals*. In his short article on *Hunting Signals as Sign System*, in which he analyzes a historical manual and music book for playing hunting signals, Karbusický proposes five different categories to understand the semiotics and functionality of these signals—or, the *sonic pattern language*—as we might put it:

1 Orders
2 Contact signals
3 Informative signals
4 Mood signals
5 Signals with heightened symbolic meaning.
(Karbusický 1986: *passim*; translated by Carla J. Maier)

Following our considerations about the situatedness of sonic signification made above, Karbusický's categories almost bluntly offer a tool for the interpretation of not only hunting signals, but the sonic signals of connected media devices in general one would be carrying around these days. First of all, Karbusický's categories work with the premise of the linguistic-lexical conventionalization of hunting signals, as well as the acoustic constraints in terms of its tones and intervals due to the limited scope of the instrument, the hunting horn. These aspects are completely congruent with functional sounds designed for mobile phones or other sonic devices. These particular affordances of the hunting horn seem essential to Karbusický's interpretation of the functionality of hunting signals:

> In this way, signals emerge as complex shapes. These aim for an optimal scope in which the communicative function manifests itself as a pragmatic-formative moment. (Karbusický 1986: 2; translated by Carla J. Maier)[1]

These attributes of hunting signals are obviously shared at large with mobile phones, kitchen appliances, or the signal of closing train doors with their limited (mostly monophonic) melodic repertoire. Significantly, the pragmatic-formative moment suggests that these signals are not a mere musical accompaniment to the hunt, but that they actually do something. For instance, as Karbusický states regarding the signal "Fire!" (*Feuern!*):

> In the signal "Firing" (No. 14) the staccato is iconically converted: it paints, in a way, the crackling of the shots, it is therefore "magically" charged. The indexical moment lies mostly in the energetic appeal; departure, activity in music are traditionally indicated by ascending sequences of notes. In No. 15 ("Stop shooting") this synergic effect is achieved by the falling octave. (Karbusický 1986: 3; translated by Carla J. Maier)[2]

Remarkably, this example addresses the whole range of functionalities and material sonic qualities that unfold with regard to one single signal: a signal that would—at first sight—be unquestionably linked to the first of Karbusický's proposed five categories, namely *orders*. However, Karbusický emphasizes that the borders between these categories are blurred, and that most of the hunting signals he analyzes are a mix of several of these five categories. On the one hand, hunting signals are bound to a very restricted 4-note scale, and the different orders and contact signals have to be learned just like a language. However, for all its linguistic-lexical quality, hunting signals are musical signals, and the corporeal and affective effects of music's iconic and indexical qualities are readily used in the design of hunting signals (see Karbusický 1986: 4). Moreover, to return to the example of the "Fire!"-signal, Karbusický mentions the discrepancy that must exist between the behavioral-communicative message and the inner feelings of those receiving the shoot-to-kill order.

The example introduced at the beginning of this section—the alarm sound bugle tune—resonates precisely with Karbusický's explication of the "Fire!"-signal: the musicality of the signal affords in both cases mood-invested interpretations and direct actions. Just as hunting signals are embedded in musical traditions, contemporary sound design for mobile phones and various other products and devices is guided by musical conventions and repertoires of sonic materials. The catchiness of a mobile ringtone lies in the juxtaposition of their signal function and popular musical phrases, and even the signals of washing machines constitute an area in which musical attributes and informative signals overlap. *Musicality*, hence, does not have to be based on obvious musical phrases or qualities, there are sonic attributes in sound design that produce catchiness through a range of other nonmusical sounds such as the sounds from car engines. The crucial aspect here is that the affective qualities of sounds are socially constructed and as dependent on

---

[1] "Auf diese Weise entstehen Signale als komplexe Gestalten. Diese streben einen optimalen Umfang an, worin sich die kommunikative Funktion als pragmatisch-gestaltbildendes Moment äußert" (Karbusický 1986: 2).

[2] "Im Signal 'Feuern' (Nr. 14) wird das staccato ikonisch umfunktioniert: es malt gewissermaßen das Prasseln der Schüsse voraus, es ist also 'magisch' geladen. Das indexikalische Moment liegt zumeist im energetischen Appell; Aufbruch, Aktivität werden in der Musik traditionell durch emporsteigende Tonfolgen indiziert. In Nr. 15 ('Aufhören zu schießen') wird diese synergische Wirkung durch die fallende Oktave erreicht" (Karbusický 1986: 3).

listening conventions as the affective qualities of music. And this can readily be extended to a range of other designed sounds such as car engines or traffic signals, as the wedding example demonstrated (see the first section of this chapter: “A sonic pattern language?”). This strong affective quality actually constitutes one major design problem that has occupied (sound) designers from the birth of the car industry (Bijsterveld et al. 2014). It is highlighted by musicologist Donald Preuss, who quoted in his book *Signalmusik* (1980) from an early description of the desired functionality of the car horn. Preuss cites *the first briefing for designing the sound of a car horn* as follows:

> In Germany a singular, easily remembered “drastic motive” was called for, with which everyone would identify the term “car,” which warns but not frightens. (“Musikalische Hupen,” 1911: 209–210, quoted in Preuss 1980: 280; translated by Carla J. Maier).

The specific model Karbusický proposes now alludes foremost to the wide variety of signals and signal music that unfolds between its function as mere commandments, and their capacity to transmit a special meaning, one that also conjures up emotions, moods, or ideological content. Moreover, Karbusický also tries to suggest a contextualized and situated interpretation of hunting signals that evoke differentiated and more nuanced listening to and interpreting of functional sounds. In studying these categorizations of Karbusický it is obvious that this—in our research quite unsuspected—historical thread of *functional hunting sounds* proves to be capable of being applied to contemporary media cultures—and their situated signification. Therefore, it is not a stretch to suggest that Karbusický’s categories reveal on the one hand the long and still quite prolific tradition of militarized sonic signs as well as contributing to an understanding of *sonic patterns* constituting a sonic pattern language. They help us in the 2010s to understand in what traditions the sonic patterns of these tiny signal sounds in our mobile devices are crafted—and they also might open up a way toward a more substantial critique of these sounds. How vivid and how adequate are these sonic patterns today, really? Do they relate to our actual practices of everyday life as of today? Or are they more remnants of earlier sonic archaeologies?

Performing sonic patterns

Returning to the example of the urban wedding ceremony from the start of this chapter, I would like to apply to this the five distinct categories proposed by Vladimír Karbusický in this final section of this chapter. I assume, that this rather experimental interpretation could help in complementing and refining the concept of sonic patterns and sonic pattern language. Karbusický’s five categories—*orders, contact signals, informative signals, mood signals, signals with heightened symbolic meaning*—reflect, again quite bluntly, exactly the range within which functional sounds operate. The example of the Turkish wedding provides with the car horn a quite specific functionality and yet a broad range of uses and misuses. According to Karbusický’s first category, *orders*, the horn can be interpreted as a direct command (“Attention! Leave the garage driveway open!”), as well as a *contact signal* (second category) to draw attention to one’s own car or to the driver, respectively. These two categories hardly tell us anything about the situatedness of the signal. The third category, *informative signal*, already reveals something about the ambiguity of the car horn sound: the informative message the sound conveys varies according to the situation, drawing attention for instance to other traffic participants or to the wedding crowd. In contrast to the others, the fourth and fifth categories (*mood signal* and *signals with heightened symbolic meaning*) are much more suited to interpretive openness: they might disclose something about the arbitrary, affective, and situated character of sonic signaling as a cultural practice. Understood as a *mood signal* (fourth category), the car horn sounds effectively turn into those musical instruments that accompanied the wedding celebration; beyond that, they gain the quality of *signals with heightened symbolic meaning* (fifth category), including the excessive performativity of gendered, ethnic, and social signification with which this sonic event is saturated (*cf.* Maier, Schneider, and Schulze 2015). So, Karbusický's categories indeed can contribute to a deeper understanding and more substantial, also social and political, critique of the various effects and ramifications of certain sounds and their in situ-usage in a given context (we will come back in the last chapter of this book, "The Sonic Capital," to the transformations of sonic labor these categories can lead sound designers to). All five categories help to stress and to focus on various sonic aspects at play and the sonic patterns developed, established, and actualized. Not the least insight is indeed the ongoing strong connection and the dispositive of the militarization of sounds and sound signals, which will be discussed to a greater extent in a later chapter (Chapter 15, especially the last section). In combination with Peirce's rematerialization of the sign process—that provides the basic understanding of signification for our interpretation—it becomes apparent, that an analytical approach to sonic signs is possible that challenges and refines a mere structuralist and less dynamic perspective on functional sounds.

Instead of affirming a prestabilized structural pattern of semiosis with its supposedly epistemological potential as a model of interpretation, one can instead engage in interpreting the specific and highly complex situated constellation of, again in Karbusický's terms *orders, information, contact, mood,* and *heightened symbolic meanings*. Functional sounds such as the car horn are situated signals, as they are grounded in temporally, spatially, and socially, almost irreproducibly specific situations of everyday life in dynamic relations to many creatures, machines, architecture, and things. With this description we follow, confirm, and exemplify, so to speak, the insights of one of the crucial treatises on sound from 1977—a time when sound studies were far from being institutionalized, even in the slightest way:

> In fact, the signification of music is far more complex. Although the value of a sound, like that of a phoneme, is determined by its relations with other sounds, it is, more than that, a relation embedded in a specific culture; the "meaning" of the musical message is expressed in global fashion, in its operationality, and not in the juxtaposed signification of each sound element. (Attali 1985: 25)

Applied to the signification of functional sounds, Attali's original claim can be further illustrated and elaborated through a very specific and playful use of a sonic pattern language, as presented by Bret McKenzie and Jemaine Clement as music-comedy-duo *Flight of the Conchords*. In their *Everyday Sounds Musical Montage* (Bobin and Waititi 2009) they combine small audio-visual snippets featuring one remarkable sonic pattern made out of household and everyday sounds: the clicking and the signal of an old digital alarm clock inspires the duo to use water glasses as chimes, to rhythmically (mis-)use light-switches, brooms, pots, sieves, and telephone push-buttons, to open and close the window to produce sonic cut-ups of the police siren emitted from the outside urban noise. The functional sounds conventionally signify in relation to the object they were designed for, and this functionality is maintained through its correct interpretation and use. In this montage, though, all the culturally embedded rules are disregarded, intended affordances are manipulated, and the sounds rigid functionality is in the end transformed. This tactic is strictly performative and challenges thus once more a purely structuralist approach to the semiotics of functional sounds. Willfully and joyfully, the existing and learned sonic patterns are dysfunctionalized and converted while performing them—in a similar way as it happened in the earlier examples with the bugle tune situation or the urban wedding.

In all of these three cases the ambiguous and conflicting sphere the listeners, the sonic consumers, or consumer citizens take up and repurpose a supposedly rigid functionality of sonic signs. The affective, spontaneous, compelling, or disconcerting effects of a signaling event are redirected to follow individual, situated needs and desires, as an intrinsic part of everyday life. In this way, in a performative reinterpretation, resignification, and restaging, these sonic patterns are indeed actualized. Following Peirce's focus on the interpreter and Karbusický's focus on the categories at play and in combination, it becomes clear that exactly this performance of sounds contributes to the existence of the patterns used. Christopher Alexander's understanding of a *pattern language* (Alexander 1977), provided us with a background of anthropologically recurring and recurrently actualized and restaged architectural situations and their materialization. This set of situations, as was apparent in the example of the urban wedding, is reappearing in situations of sonic signals and sonic patterns, stressing the complex interweaving of haptic, visual, kinaesthetic, and sonic experiences in one's individual, phenomenological perception of everyday life. Stable and established *architectural design patterns* of everyday life imply almost, so to speak, the actualization and performance of more dynamic, *sonic design patterns* and the establishing of a relevant, language-like structure of these patterns. Sonic patterns emerge out of the situated character of the actions taking place in design patterns. This insight, though, can only be articulated by us in the framework of a *sonic materialism* that grounds and motivates our research. In this interpretation, the situation of a sonic signal, the context, the highly temporalized, spatialized, the culturalized and corporealized character of this specific sensory experience is explicated by all the material, physical, experiential, also historical, semiotic, psychological, and phantasmagorical threads in this one sonic experience. It is this sonic materialism that guides us—as *interpreters* in all the many meanings of this word—performatively to actualize sonic signals in a given situated context. Sonic patterns are therefore actualized and solidifed as a language-like set foremost by means of performing and by playing with them. This performance of sonic patterns takes place in an orderly use and willfully disorderly, comic and joyful misuse and it is not necessarily tied to an audible reinterpretation. The reinterpretation and the actualization take place with the same impact and meaning as reactions to and actions escorting and resulting from sounds. Listening and sounding are here strongly interdependent activities of a *sensory corpus* and at times also for a *sonic consumer*: "Dancing is a way of hearing; singing is a way of dancing: Singing is a way of hearing" (Schulze 2018: 231). A *sonic pattern language*, hence, exists as materialized as soon as it is possible to misuse its implicit rules, material characteristics, and performative guidelines—and as soon as this misuse can be recognized, sanctioned or enjoyed as such. It is the domestication of these sounds that reinforces their underlying sonic patterns. Signifying sounds can be understood as such a performance of sonic patterns.

10

Situated Signaling

Sound practices in personal situations

Singing the signals

I'm sitting in the ICE, a high-speed train, from Berlin to Munich. Having breakfast at the onboard restaurant, I become the witness of a spontaneous act of sonic imitation and appropriation of functional sounds. When the train is standing at the gate of a train station and is about to depart, I hear the usual *beep beep beep* of the door signal resounding. The restaurant's attendant who has been in a jovial mood throughout the journey, starts to cheerfully imitate the beeping sound three times in the exact pitch and longitude, and then adds his own, improvised melodic line, *dada da daa daa da*. This little interlude actually sounds very natural and as if done without much intention. When I ask him what the song was that he has just sung, he looks at me somewhat astonished, and says that he has absolutely no clue. He adds that it just happened and that he was actually not aware that he was singing a particular melody at all. I say that I recognize the door signal in his song, he is amused and even more surprised of his own unawareness. Asking him about his musical preferences, he responds that he listens to "a bit of everything." I associate some sort of folk music tune with his spontaneous song and, influenced by his slight Bavarian accent, combine that to suggest it might have been a Bavarian folk tune that harmonizes so smoothly with the door signal. To this he turns away rather brusquely saying "well, no…!"

This small and rather easily forgotten episode while traveling, commuting, and working exemplifies quite nicely how functional sounds are individually and spontaneously reinterpreted and reframed in everyday situations. A form of appropriated reproduction takes place here, a bit of improvisation also, and, obviously, there is no major intent to be found here. There are, though, a series of striking aspects of how sounds occur that one can observe in this episode. As in many possible instances of living with sounds, they occur in a specific time of the day, of the week, in the year, and at a specific place with specific social activities, individual trajectories, memories, and desired or feared outcomes. It is actually quite hard to transfer this very example to another time of the year when for instance clothes are thicker, people are more cocooned in their places or into another nation state or culture where maybe such practices are either very common or considered very inappropriate if not illegal. This represents the main characteristic of situatedness that indeed plays out here in the most striking way.

This *situatedness*, however, has various consequences for the actual sonic performativity, for the domestication of these sounds, their sonic pattern, and the relation of these sounds to the listening and resonating, the singing and sounding individual. First of all, and already implied in what was said here earlier, this situatedness of such sonic signals is individual: it is also not easy to transfer this minuscule sound performance of the train's attendant to any other person or official in this or another train or in another vehicle of public transport, a tramway, subway, a helicopter, or an airplane. This individual character is, though, not arbitrary but specific: it characterizes exactly *this very situation*, its time and location, its personnel, its material objects, and its manifest performative acts. The situatedness of these sounds and their appropriation and assimilation via singing is individual in all its intimate, oblivious, and sometimes even gory details. This individuality of sonic experiences and sonic domestication, though, is again contrasted by the collective character of such situations. On the one side, such situations are most of the time in our lives as humanoids not happening in complete seclusion far away from any other humanoid alien. Sounds are, most of the time in known cultures, projected and performed in situations of at least possible sociality and surely of a mediated sociality as sounds resonate through walls and any visual obstacle that shields us from the view of others. This is how a sonic pattern language is established. The *aural arena* in which sounds occur might be limited to one apartment; one office or cubicle; one restaurant, bar, or club; or one train, wagon, or car, but the *acoustic horizon* will surely extend and transcend the obvious visual and haptic limits of this situation (Blesser and Salter 2007). You can always hear the mumbling, screaming, or typing next door, the toilet noises in the next stall, or the sounds of munching, drinking, laughing, flirting, or nagging at the neighboring table. One just pretends to not listen, to not hear, to not register what is going on all the time in a shared social situation—even if it is delimited by walls, windows, curtains, or screens. The reason for this lies, on the other side, in many shared practices, inclinations, desires, and intimate routines, fears, or little pleasures we indulge in the everyday. Joining into a song, moving your legs or hands or head in resonance is not a surprising activity for humanoids. Though the individually apt practices are differing across cultures and historical eras, the mere fact of *bodily resonance* and a *joy in repetition* and joining in seems not to be unusual at least for the humanoid aliens known in history, in the present time, and in fiction. The assimilation of sonic signals is a collective practice.

It is exactly this situated interweaving and intermingling of collective practices and relationality with individual habits and desires that also gives way to surprising, highly randomized, and unexpected encounters: chance occurrences intervene in these situated collective-individual sound environments. Now, the degree of randomization of any activity or intervention in a situation is harder to tell than it seems (Schulze 2000): one overestimates easily the random character as soon as the structuring formants are complexly layered enough. As cultures and individuals generally do not operate according to chance operations, and as these operations are also occurring in certain sequences, clusters, and rhythms tied to preexisting frames of weather, work life, family life, cultural traditions, and basic needs, the room for radically aleatoric actions can often appear as tiny as can be. Yet, surprising encounters do take place and they rarely are not taken as a generative nucleus for further actions: it is precisely this rare nature of chance occurances that makes them seem so attractive, extraordinary, and almost necessary to connect with. Another side-effect of this situatedness is the aptness to dysfunctionalize the heard sounds. As sounds propagate and resonate, they can be taken up—and also in unusal, unexpected, and truly unintended ways: performativity here gives the opportunity for an instantaneous and spontaneous resignification of any sound. One can dance to car alarm sounds (LaBelle 2010) or mobile phone alarm sounds (*cf.* Chapter 9, section "The semiotics of hunting signals"), one can compose with system sounds (SomethingUnreal 2007), one can songify and harmonize the erratic ramblings of a person under anesthesia (Publio Delgado 2016), and, obviously, one can also sing with door signals on a high-speed train. The domestication and appropriation of sonic signals and sonic patterns is performed in exactly those excessively and joyfully dysfunctional and intended abuses.

This little, if you will, holographic analysis of this minor situation of sonic signals and its domestication, from many various perspectives and aspects indicates the everyday tactics of the sonic consumer—tactics in the sense of Marcel de Certeau's famous definition in *The Practice of Everyday Life*:

> The space of a tactic is the space of the other. Thus, it must play on and with a terrain imposed on it and organized by the law of a foreign power. … It does not, therefore, have the options of planning general strategy and viewing the adversary as a whole within a district, visible, and objectifiable space. It operates in isolated actions, blow by blow. … It must vigilantly make use of the cracks that particular conjunctions open in the surveillance of the proprietary power. It poaches in them. It creates surprises in them. It can be where it is least expected. It is a guileful ruse. In short, a tactic is an art of the weak. (Certeau 1984: 37)

In the *sonic tactics* of the singer, the dancer, the mixer, and the author of an *impromptu collage* the appropriation of a sonic pattern takes place. It takes place, though, not on an abstract plane but in a specific situation that is the ground and the arena for tactics. As Sophia Prinz states in her article on *Alltagskultur als Designkultur* (*Everyday Culture as a Design Culture*), the incesssant negotiation between the world of objects—the so-called *Dingwelt* (thingworld)—and the individual manifold *corporeal practices* generates a wide variety of intentional and nonintentional tactical shifts all the time. This ongoing process of performing (sonic) design patterns through their (sonic) affordances results in a domestication that operates mainly by way of dysfunctionalizing an existing affordance:

> It has been pointed out from various angles that many of the prefabricated artifacts and devices are not only used for their original purpose, but are reused, transformed, and transferred into new contexts—be it the bottle of wine used as a candle holder, the cell phone, which is decorated with stickers and glittering stones, or the 1980s cash-glasses, which for a few years now decorate the noses of Berlin hipsters. (Prinz 2013: 41; translated by Carla J. Maier)[1]

This process of appropriating, dysfunctionalizing, domesticating, and intentionally misusing is never limited to materially stable objects. Sonic signals—even the most bland ones— can become highly personalized as well, even with a taint of unsettling intimacy and daunting memory that makes you shiver. On a sunny weekend, my husband and I went on a trip

[1]"So wurde von verschiedener Seite darauf hingewiesen, dass viele der vorfabrizierten Artefakte und Geräte nicht nur ihrem eigentlichen Zweck zugeführt werden, sondern je nach Bedarf umgenutzt, umgestaltet und in neue Zusammenhänge transferiert werden—sei es die Weinflasche, die als Kerzenhalter dient, das Handy, das mit Aufklebern und Glitzersteinen dekoriert wird, oder die 80er-Jahre-Kassenbrille, die seit ein paar Jahren die Nasen von Berliner Hipstern ziert " (Prinz 2013: 41).

along the Rhine river in Germany. As we walked along the waterside promenade and passed a railroad crossing, the sonic signal of the closing barriers took my breath away, made my mouth feel dry and my muscles stiff, and I was overwhelmed by a feeling of being displaced. The sound immediately reminded me of the devices and their various sounds at an intensive care unit controlling the heartbeats of my father, who had been put in an artificial coma after suffering from a severe stroke. I recalled the whole situation, my anxiety, my hope, my helplessness—just by being exposed to these similar sounds, their rhythm, their pitch, their sonic shapes. I performed these sonic patterns in an unusual, very personal way. These rather unremarkable traffic signals teleported my sensory corpus swiftly back into this harsh moment—ten years ago, in a clinic and in a room I haven't been to again since those days in 2007. I had not even to sing those signals myself; *they sang the memory of my father in me.* My body remembered these sounds and what they signaled. Remembering functional sounds and their situated meaning factually dysfunctionalizes them and reassigns them a tactically new and very personal if not intimate meaning.

Situated meaning

As we saw, the individual experience and interpretation of sound events and sound signals is strongly related to a specific situation. But how can we conceptualize this specific relation in the framework of semiotics: between, on the one hand, the corporeal, experiential context of a situation in which one operates with signs and, on the other hand, how sound can be experienced as a signifying signal? To a large extent, the existing research and literature on semiotics does not primarily reflect corporeal or situative aspects of semiosis. And in many aspects of operating with signs in daily life and professional use this might not be necessary. But it seems to be urgent as soon as *sound* is interpreted as a signifier. In our fieldwork, we observed this fact especially regarding the work environments of sound designers, the design teams and client relations, as well as in the ensemble work in the studio or in front of the main production device. The previous chapters in the part on "Sonic Labor" presented various examples of this corporeal and situated substance of sound production and listening to sound. It laid out the specific materiality and situatedness of sound invention, sound production, and sound design.

The problem we are confronted with here is a fundamental difficulty in many contexts in which a semiotic process has to deal with highly dynamic, transformative, and situative artifacts: *How can one understand meaning if one cannot simply refer to a corpus of definitory and semantic descriptions and code-mappings? How can one understand that which seems to evade the known universe of signifiers and signifying relations?* The established notion of semiotics apparently reaches its limits exactly with such dynamic artifacts. Hence, in understanding the age-old and factually still unanswered question *how sonic events, how sounds and even music might become meaningful* and how humanoid aliens like you and me attach specific meanings to specific sounds one would need an approach to meaning that would rely more on the situated context and the individual corporeal experience of certain sounds than to a general, immaterial definition thereof. In the framework of a larger anthropology of the senses the status of a situation can be described as follows:

> A humanoid's body is *never without* a situation. An alien's experience is *never not* situated. At a given moment one might decide to leave the characteristics of an actual situation aside and exchange them for other, maybe imaginary, sensorial, unconscious, abstract characteristics;

> then again one finds her- or himself in yet another situation: in an imaginary, a sensorial, an unconscious, an abstract one. The individual body is situated and it is located. It is characterized by its temporal and environmental situation. (Schulze 2018: 140)

Concerning this relation between the sensorial, the sonic, and the body there is a phenomenological and psychological approach which might help here in understanding this highly sensorially and corporeally anchored relation. I am speaking here of the research on semiogenesis and pragmatic generativity as proposed by Eugene T. Gendlin. The American philosopher, phenomenologist, and psychologist uses a concept of generativity that includes embodied but not yet explicated meanings. The term he uses for this concept is *felt sense*. The term *felt sense* refers to the permanent grounding of one's perception and activities in a large number of proprioceptive sensory events. You or I, we feel ourselves in our bodies, in situations, among other people, and with certain constellations of expectations, obsessions, and fears. But we actually do very rarely explicitly *interpret* these continuous sensations, these ongoing feelings; only in certain, more crucial instances we then suddenly do. A stunning example Gendlin gives is the following:

> For example, suppose you were at a party and felt you were bored and needed to go home. But suppose that instead of going home, you opened up the boredom and found anger. And suppose that in finding the anger you found also that you needed to stay and say something directly about the anger to someone. In a similar way as we pursue a goal, the goal seems to change. But later we say the new goal is the one we really wanted all along but we didn't know it. (Gendlin 1992: 203)

A person in question is in this example drawn back from their routines of perception, of interpretation, and even of feeling to a hightened attention concerning their actual needs, desires, wishes, feelings, and lacks. In this example, it is the *boredom* that has a generative quality. But anything might work as a hint to focus onto one's own feelings, interpretations, and perceptions in a given situation. Following Gendlin it is not so much relevant *what* points us at this task, but the mere fact *that* we are alerted and we are subsequently focusing on this. These alerts to a new focus are led by practices and by specific empirical situations. Or as Gendlin puts it: "Any situation, any bit of practice, implies much more than has ever been said" (Gendlin 1992: 201). The amount of intricate implications that are actually necessary to understand the many details of a given situation, can seem endless and almost inexhaustible. With special regard to sound design, sound performances, or sound inventions it can indeed seem that the role a sound, a sonic pattern, or a sonic affordance plays in our lives is much less defined by its acoustic or electroacoustic properties than by its individually assigned situated role in their, your, or my specific life, right now, right here. Meaning—according to this corporeally anchored theory of signs by Eugene T. Gendlin—is generated, manifested, and stabilized in a corporeal and experiential situation with its specific and often unforeseeable long-time dynamics. Meaning is not stored, mapped, and retrieved—meaning *comes*:

> This coming is characteristic of the body. What else comes like that? Sleep comes like that, and appetites. If they don't come, you just have to wait. We all know that. Tears come like that, and orgasm. Emotions come like that, and so also this felt sense, which is wider and at first not clear, comes like that. (Gendlin 1992: 194)

What Gendlin calls the *felt sense* (Schulze 2018: 145–150) is exactly this corporeally anchored sense for the meaningful character of a situation: it is its *situated meaning*. Hence, a semiogenesis is inextricably bound and dependent on a given situated constellation of sounding and listening, a nexus between oscillation and resonance. A new meaning of a familiar or unfamiliar sound is therefore contingent in a sonic and sensory situation:

> You can feign joy or anger but to have them, they must come. So also does the muse come, when she is willing and not otherwise. And new ideas, *the lines of a new design*, and steps of therapy come this way. (Gendlin 1992: 202; emphasis added)

The *lines* of a new design—or the *sounds* of a new design—emerge from such situated constellations. Situated constellations like those we encountered recurrently in our fieldwork—be it in a project developing sounds for endoscopic machines of an intensive care unit, sounds for the branding of a new development quarter in a metropole city, or the sound design of a museum's audio guide. In all these cases the corporeal outlines of a new performance of an old sonic pattern actually generated the outlines of a thoroughly new sonic pattern. With the generation of new meanings also new designs were generated.

Sonic technologies of the self

In her article, *Music as a Technology of the Self*, Tia DeNora explicates her approach to understanding the roles music can play in today's everyday lives. This notion, though, of a *technology of the self* extends apparently into the field of functional sounds and their everyday use:

> Music is a material that actors use to elaborate, to fill out and fill in, to themselves and to others, modes of aesthetic agency and, with it, subjective stances and identities. This, then, is what should be meant when we speak of the "cultural construction of subjectivity"—and this is much more than an idea that culture underwrites generic structures of feeling or aesthetic agency as is implied in so many post-structuralist writings and by musicologists trained in semiotic analysis of texts. (DeNora 2000: 74)

It is important for our understanding of functional sounds to acknowledge that the *cultural construction of subjectivity* is not an abstract or invisible or somehow magical operation taking place without individuals, without listeners or designers or prosumers actively involved. The contrary is the case: it is the actual, the everyday, the very material and physical activities of *doing things with sound*, of *ringtone acts*, or of *material sound gadget culture* that construct who you or I or they are. DeNora's approach can be interpreted here in such a way that her statement on musical experience and its generativity in individual lives and biographies of subjectivity and of experience extends to the whole realm of those tiny, miniaturized microsounds that punctuate our daily lives. Hence, we would like to add: *sound is a material that actors use to elaborate, to fill out and fill in, to themselves and to others, modes of aesthetic agency and, with it, subjective stances and identities*. The situation in which this takes place—many times on every single day—is not only generative in itself but it is crucial to the whole experience and usage of functional sounds.

What DeNora does for the analysis of music is relevant for the analysis of functional sounds in everyday life, as it contributes to refining the significance of sound in the cultural construction of subjectivity. Within discourses of cultural semiotics, however, the link between sign systems and concrete cultural practices often remains somehow awkwardly formalistic, as can be observed in semiotician Roland Posner's semiotics of a pop concert in *Basic Tasks of Cultural Semiotics* (Posner 2004): while the complex relationship between different sensory modalities, matter (air) and the technical apparatus (spotlights and projection screens as well as musical instruments, microphones, and amplifiers), the involved social institutions and the textual and linguistic codes that are at work at a pop concert is highlighted, the allegedly unambiguous meanings that can be attributed to the event seem too readily fall into place:

> This special constellation of media predisposes it for an emotionally-laden, *generally understandable message*, which can provide every individual in a large audience with a *feeling of belonging*. (Posner 2004: 63; emphasis in the original)

This analysis seems to imply that the meaning of a sonic event such as a concert is derived directly and unmistakably from the technical, visual, and verbal components as such. As if they *were* stable and ever unaltered messages that naturally emanate from loudspeakers, verbal exclamations, and bodies. As if a pop concert belongs to an unambiguous sign system that works for every pop concert in the same way. What this analysis fails to ask is: *Where* did the concert take place (a huge concert hall or a small backyard club)? *When* did it take place (in the 1950s or in the 2010s)? *Who* was present? *What* were the specific sounds, movements, relationships, interactions, hierarchies? *What* are the specific sonic patterns that signify in regard to this material, situative, and corporeal sonic event? *How* is such a "generally understandable message" constructed, and for whom, and *who* might be excluded because they do not understand, or are not meant to, or do not identify with the content of this message? *How* is a "feeling of belonging" created, *what* are the sonic, social, individual parameters of belonging, and *how* does the feeling of belonging, or not-belonging to speak in Gendlin's terms, "come to you"?

A pop concert, hence, is such a technology of the self as explicated by DeNora. Yet, such a technology is never of general relevance for all humanoid aliens or sonic consumers at all times. As was elaborated earlier, sonic patterns do not work for everyone, and everywhere, in the same way. Moreover, they demand a performative and corporeal actualization and embodiment to consitute an integral part of a social situation. There are, for instance, a large number of highly culturally and nationally defined sonic pattern languages in the area of sports activities and sports media events. And in these events there are also continuously shifting relationships between different sounds, different sonic patterns, different performances of sonic patterns, and the different ways in which they signify. The sounds that are, for example, characteristic of a soccer game in a big stadium in Germany might be familiar to many people who live in that country—being dedicated football fans or not. But rather than representing a national code, the sonic patterns at work here represent a major reference system for the soccer community in that country and all fan cultures attached to it (*cf.* Bonz 2015). But even within this community there are different sonic patterns at work which both mark and challenge significations regarding ethnicity, gender and class, region and city. While the collective power of cheers and songs in soccer stadiums seems too often to work as a tool for reassuring a feeling of national collectivity, it has at the same time an exclusionary and dividing power that redraws boundaries toward other nationalities, or continues to justify an agreed concealment of openly lived homosexuality. Hence, it is quite complicated to extract or to define a "generally understandable message" here. Therefore, the sounds in a soccer stadium might be understood also as sonic technologies of the self.

Another example that stresses the situated politics and shared aesthetics of sound patterns can be found in the sonic materiality of skateboarding (Maier 2016). Understood by some people as a nuisance that is to be banned from public spaces within the city, the particular sounds that emanate from a skateboard grinding the urban concrete are immediately deciphered and invested with meaning by members of the skateboarding community. These sounds constitute sonic patterns. The sounds of skateboard wheels as they grind the surface of concrete ledges, and the ringing and scratching sound when skateboard trucks hit the handrail during a railslide—go straight to the guts of skateboarders to whom these sounds resonate with hours and hours of practice, of fun and frustration, of overcoming the fear of getting hurt, and the pride felt when being cheered at for finally standing a new trick (Maier 2016). All of these performances of sonic patterns at a pop concert, in a soccer stadium, and whilst skateboarding can be recovered and discussed in the working environments of sound designers. Whereas these three examples of collective performances of sonic signals are established, learned, and performed over time, the work in sound design exactly connects to these traditions and intends to develop those, to add certain signals or to substract others—for example, in our fieldwork, when crafting the soundtrack and the sound design of a non-Blockbuster, mockumentary or when designing the sounds for a newly developed casual game on various smartphone platforms. Hence, in the process of designing sound a sonic pattern language with specific references is very often established (as presented in Part Two "Sonic Labor," Chapter 6 on the "Skills and Habits" of sound designers, especially in the section "Make it fantasy!"). In designing sound or designing new sonic patterns a new performative practice with these sounds shall be inspired—and therefore the description of these sounds reaches incredibly detailed aspects of a sound signal. While determining a new sonic technology of the self the established technologies in sonic patterns are reviewed and adapted, developed further, maybe dysfunctionalized. The highly specialized language or jargon that is established between the different actors involved in a sound design process intends to communicate in the most precise way what kind of sound is meant to be created and when a sound actually "sounds right." This working language on sonic materialities varies obviously according to the musical and media socialization of the designers as well as according to the level of familiarity and friendship between the colleagues. It also can change drastically between different projects with mainly the same team but with a different client and maybe one or two new team members.

Sonic patterns and their materialities can then be described by professionalized terms such as "arcade sound," or highly individualized expressions such as "The Joe-Voice" or the reference to one famous composer or sound designer be it Brian Eno or Hans Zimmer (*cf*. Chapter 6, section "Make it fantasy!"). This terminology, this highly individual working language to describe sonic patterns and materialities, is mainly established between the colleagues who have agreed to use this one term to refer to one specific sound design. Yet, it is not impossible that such a term and notion can catch on with new team members in a sound production and then even be mentioned in more public situations of presenting or discussing sound design—as for instance this academic study you are reading right now on the cultural theory of sound design.

Sonic technologies of the self are therefore on the one hand highly individual, yet they can on the other hand catch on very easily with others. Then they can become a trend, a hype, a fashionable fad, or even a common or widely accepted cultural practice. A sonic pattern, its sonic affordances and its preferred performance are then established. This establishing, this implementing of a new sonic affordance or sonic pattern can surely be called the goal of most of the sound design projects we encountered in our study.

This goal then is easily recognizable in all its paradoxical and highly dynamic aspects: if you intend to inspire individual, personal, and situated uses and misuses, performances of a sonic pattern you proposed, then the *nonintentional* misuse in performing, in appropriating can be as probable as an orderly use. The process of domesticating sounds is, therefore, the ally of all sound designers and at the same time it is also their archenemy. A sound designer wishes for this kind of massive misuse—being a public sign of ultimate recognition—but they also fear it, as the originally intended use might then also vanish in the haze of this hype.

Domesticating sounds

Toward the end of this chapter we would like to go back to the starting example of this chapter: humming the functional sounds of a high-speed train right on this very high-speed train. Perceiving and living with sound is, as we saw, a performative act with a situated and sensory character that constitutes one of the technologies of the self. To such an example you might add various other experiences, assimilations, and domestications of sound signals or functional sounds given to you by a certain work environment, by computer networks, or by a mobile gadget you are carrying around with you. In all of these cases one experiences how sound, a *sonic environment* or a *sonic artifact*, is being appropriated. This process of bodily and habitual appropriation applies several specific means to domesticate such a sound—and with it, presumably, a whole cultural environment, a technological dispositive, a recurring situation of everyday life. With these thoughts we follow, again, Roger Silverstone's theory of *cultural domestication of new media and technology*. Earlier in this book (Chapter 1, section "Functionality as guidance") we translated the main research question of domestication theory tentatively into the area of sound studies: *does the domestication of new products involve, quite literally, a taming of the wild and a cultivation of the tame by means of sound design?* Now, nine chapters later we could rephrase this translation with an even stricter focus on the role of sound design itself: *does the domestication of new and wild sound designs also involve, quite literally, a taming of the wild and a cultivation of the tame?* Or, more generally: how is it possible at all that one is listening to a designed sonic environment—maybe starting out with being surprised or even shocked by some sound events—and one is nevertheless arriving at some sort of *understanding, appropriation of, engagment in* this new, functional sonic artifact? How does this process of *domesticating sounds* actually take place?

At this point it might prove helpful to add a situation of sonic labor we encountered in our research to the exemplary situation of someone singing the functional sounds of a high-speed train: the situation when a sound designer (or a team of sound designers, musicians, sound engineers, and developers) has to decide upon what sound to use for a specific function in a newly built transport vehicle. It is safe to say that—also referring to the findings displayed in the previous part on "Sonic Labor"—such a decision is not made in a neatly organized process. In the common back and forth of project management in many areas of *sensory* labor, a decision like this is more forthcoming to its actors, they are serendipitious but have conceptual guidance and use self-reflection in the process of creation when approximately approaching the final version of the sound design. There is not one point when this decision is willfully made; the point to end this search process is just suddenly made—often though externally forced by hard deadlines, end of budget, or client demands. The agglomeration of activities, of revision loops and harsh turnarounds, of newly introduced major directives for designing the desired artifact and also of newly introduced budget limitations or unexpected budget expansions—all of this (as layed out visually in Chapter 7, especially section "Pitching detour") contributes to an at times irritating uncertainty and at other times to a rather generative incentive while designing. In order to understand such situations of decision in the design process they can be interpreted as first manifestations of a larger process of domesticating sounds *by the very designers of these sounds*: when—referring to our rephrasing of Silverstone's dictum—not only *wild* sounds are tamed yet also *wild* designers and their joined taming presumably is intended to result in *a cultivation of their tamed* sonic artifact. This process of domesticating certain wild sounds and proposed designs does apparently find its correspondent taming activity in the playful assimilations by listeners, users, and everyday clients of such commodified sonic artifacts. If one would follow this path then almost all sonic performativity could soon be subsumed very convincingly as an example for various processes of domestication—be it related to functional sounds, to canonized musical pieces, to learning a musical instrument, to composing a playlist, or any other possible forms of applying sonic technologies of the self, of *musicking* (Smalls 1998). Musicking can be a form of sonic domestication.

At this point in our investigation the inextricable coalescence between *sonic labor* and *sonic consumption* becomes overly obvious. Any sound designer in any state of their work on a sonic artifact or sonic pattern is incessantly also a sonic consumer at the same time, proceeding to domesticate certain present sounds still perceived and experienced as *wild*. Hence, the contemporary audible shape of various artifacts, of gadgets, of soft- and hardware, of instruments and online applications used to invent, to produce, to postproduce, to master, and finally to distribute and to advertise any sonic artifact, this sonic shape of sonic labor itself, is a labor condition with a supposedly not too minor impact on the artifacts produced. While enveloped in sound one unravels ideas for sound: this is the *prosumption*-paradox. A sonic consumer of sonic artifacts is—in many cases—also involved in sonic labor.

How does this insight now alter our approach toward sound design or toward functional sounds in general as soon as we take a process of domestication as crucial for any activity of such musicking or *sonicking*? Such a major assumption has the potential to turn any thinkable cultural theory of sound design upside down: it could reverse the responsibilities and the effects of sound production; it might install the user actually as a major actor in this process of domestication instead of the producer; it might also stress the pervasive and ubiquitous character of sound designs to an extent that supposedly negligible situations of sound design usage turn actually into exemplary samples of a much larger phenomenon; it might also point to the fact that the process of semiogenesis, the generation of meaning regarding certain sounds (already discussed in Chapter 9), definitely *never* actually follows the well-known diagrams of mediated transmission reduced to a *transmitter-message-receiver*-relation. What relation of generating and transmitting meaning of, through, and around sound are then instead confirmed by this new approach of domesticating sounds by sonicking?

If one would indeed propose such a complex notion of generating meaning then one would finally leave behind all the simplifying interpretations of semiotics; in contrast one would then go back to the vibrant first writings about semiotics from the nineteenth and twentieth centuries as published by Charles Sanders Peirce, by Ferdinand Saussure, by Max Bense, or even by Umberto Eco. All of these authors actually stressed the material and at the same time the reciprocal quality of sign processes. *Semiosis*, the process of generating and thus using and understanding signs has been defined by all these thinkers of semiotics as a process in *material* substances that is not superimposed on nonresponding substances but in contrarily developed by users and things, interpreters and materials in a joined activity of sign operations (Chapter 9, section "Rematerializing sonic signs" provided the initial argument to this conclusion). Bluntly said: following these authors and stressing the performative and situated character of signifying, especially while using sounds, one must acknowledge that there definitely is no *immaterial* way of transmitting meaning that does not affect the people involved—as well as that there might never exist a meaningless use of material substances that won't affect people involved. Signaling is situated, material, and affective as well as generative if not transformative. In this loop of assigning meaning while using and stabilizing meaning while recurrently being used, being heard, being sounded—in this form of a continuous repetition the domestication of a sound can take place. It is this reiteration of *sound operations* that constitutes a sonic experience. Hence, it leaves sonic traces on your skin, muscles, and bones, your eardrum, your various diaphragms, in your stomach and cranium. Repetition effectively inscribes the reiterated sounds into one's experiential sensory corpus. The sound becomes visceral and manifest. The sonic experience is established and implemented. A personal situation, hence, provided through sound practices the actual process of signal transmission.

11

Transmitting Sounds

Toward a sonic mediology

To transmit and to receive

The concept of transmission, or *la transmission*, takes a central place in the French cultural historian's Régis Debrays's outlines of what he calls a *médiologie générale* (Debray 2000)—an allusion to famous introductions to a *sémiologie générale* or a *sociologie générale*. Yet, this new kind of transmission theory must not be confused with theories on transduction, especially signal transduction, as they are known in the field of acoustics. The form of transmission Debray has in mind focuses on exactly one detail that is lacking in most modern theories of transmission. Or, as Michel Serres stated as early as in 1985:

> Transmission trumps listening, we know how to project a sound and how it spreads, we know how to relay it, we are not good at receiving. (Serres 2008: 139)

Serres and Debray agree on one diagnosis regarding science history: the great white unknown territory in roughly all theories on transmission is the part of reception. Projection is regarded as a major issue, reception is taken for granted. Following the distinctly linear logic of transduction the activity of receiving is mainly modeled as a negative or an inverted activity matching previously executed processes of sending or projection. Hence, the only possible role left for receiving in such a dispositive of transmission is the role of a dependent, a heteronomous, and necessary side-effect—maybe with selectively innovative interpretations and appropriations if one includes cultural theory approaches to coding and decoding (e.g. Hall 1973). Receiving has no dignity as an activity in itself and it is of no real relevance in traditional theories of transmission. Dialectically, this provides a clear work assignment for this cultural theory of sound: *Is it thinkable, that the activity of listening, of receiving can retain a genuine, a central, maybe even an unexpectedly active part in a concept of transmission?* Could one conceptualize a form of transmission *without* relying exclusively on models of signal transduction that govern models of transmission in acoustics as well as in quantitative research on media reception and on audience reception? Is there a way of explicating a process of *cultural transmission* with a strong respect for the process of *cultural reception* or *cultural appropriation*, for *assimilation* and *domestication*? Can there be a model of transmission that does not isolate and therefore falsely idealize the role of the sender, but maybe a model that would consider most of the actors in a nexus of transmission as *receivers*?

Certain traces for such a model one can find in the works by Debray, hence, in this chapter I would like to explore and to translate it into our research. First of all, Debray does propose to use the french term of *transmettre* as a verb and as a concept that goes beyond the known focus on *communication* in media research. Therefore, he does not ontologize transmission as a fundamental a priori to any human interaction but as an intentional, a performative, and an anthropologically specific, historically and culturally anchored activity that humanoid aliens choose to undertake. Thus, *to transmit*—following Debray—is not a basic condition of any humanoid alien alive in sufficiently organized societies but a highly refined and cultivated form of operation: it requires arduous, trained and exercised, taught and learned, spohisticated and erudite operations. It is part of an elaborate material culture as opposed to the immaterial culture that—in Debray's reading—is implied in the cryptoreligious if not cryptocatholic term of *communication.* To put the focus on communication as a main paradigm of speaking about culture according to Debray moves into the empire of the immaterial as it locates a process of transmission not firstly in material conditions, prerequisites, objects, tools, skills, and sequences of actions but in a rather superhuman, a spiritual and mental condition, a semantic strength of a message solely to be found in a strong and able sender personality. Yet again, like in the rematerializations of semiogenesis (as discussed in Chapters 9 and 10: "Signifying Sounds" and "Situated Signalling"), here an *immaterial* transmission is also unthinkable—not only in the Marxist terms Debray had been educated in.

The material transmission of goods, of products, and of *materialized ideas* is consequentially the main concern of Debray's mediology. His fundamental question is: *how do materialized ideas travel through time—and how do they succeed in moving a large mass of people?* This outline of *the mediological project* in one research question makes it quite clear: Debray conceptualized his theoretical approach as a complementary counterpart and a truly necessary continuation of the *sociological* project and of contemporary communication research. In his vision the research framework of mediology would generate in the near future the same vast amount of academic efforts and the same research impact as sociology had since its impressive foundation—for instance with the writings by Auguste Comte in the nineteenth century. Apparently, this has not happened yet (at least not until handing this proofread manuscript to the publisher). But what did happen in the last decades is still a quite impressive and undeniable refocusing on questions of media studies—and recently also on the *media a priori* as proposed by Friedrich Kittler and his approach of using literary studies to research issues from science and technology studies. Debray's focus on media though is thoroughly different, if not in direct opposition to that of Kittler. Whereas Kittler's approach focuses actually still on interpersonal processes of transmission that are guided, taken over, dislocated, and deterritorialized by various forms of machines, of technological inventions, and of tools, Debray leaves the whole area of interpersonal communication behind. Mediology does not focus on the transmission of character strings or how they might be subsequently processed, altered, newly sculpted, scripted, or thoroughly reversed but on what Debray calls the *cultural transmission*: "cultural transmission begins where interpersonal communication ends" (Debray 2000: 98). Whereas this brief notion might be mistaken for an immaterial-spiritualist idea of transmission, let me just add this dictum by Debray:

> to transmit is to organize, and to organize is to hierarchize. Hence also to exclude and subordinate—necessarily. (Debray 1996: 46)

By all means, Debray's approach to transmission is a materialist approach (Spoiden 2007). He even goes so far as to state that transmission must be understood as a physical and as such intrinsically transformative form of *transport*:

> Transport by is transformation of. That which is transported is remodelled, refigured, and metabolised by its transit. The receiver finds a different letter from the one its sender placed in the mailbox. ... To transmit should not be considered merely to transfer. (Debray 2000: 27)

Here, the materialist and also the historical aspect becomes obvious. Conventionally, one would assume that a message that is transmitted physically could retain its essential qualities and that exactly this stability of properties qualifies as transmission. Debray though denies this assumption and stresses that as soon as any object (containing maybe a form of message) is transmitted over time, into other territories, other cultures, other historical eras, and into radically differing contextual environments, the transported object is irrevocably altered, substantially. This alteration, however, is basically neither in the doing of a sender using nor of an engineer developing this apparatus for transmission. It is a material property of any actual transport. As Steven Maras points out:

> Debray's discussion is not focused on the sender as abstract actor, but on the intellectual as defined and constructed under particular material conditions. (Maras 2008)

In contrast, Debray's concept of *cultural transmission* stresses exactly all the contested, the unstable, and the heteronomous qualities of transporting. Maras summarizes these qualities:

> "transmission is a *historical process*," defined by a "thick temporality" ... "transmission is a *collective process*," involving many people on the line ... "transmission is a *violent collective process.*" Transmission on this understanding is a combat against noise, inertia, other addressees, other transmitters. (Maras 2008)

As a *historical, collective*, and *violent* process transmission is therefore an activity that is relational, transformative, and multicentered, undergoing various troubles, obstacles, and hindrances, manifold instances of resistances that need to be acknowledged, analyzed, and tactically overcome. This complex account of transmission apparently transcends most of the hegemonic definitions of transmission, signal transduction, and signal processing these days. The focus of Debray's definition lies—following this interpretation by Steven Maras—neither on encoding and decoding (as in Claude Shannon's model; Shannon 1948) nor on the decoupling of both (as in Stewart Hall's model; Hall 1973) nor on the necessity of a common code (as in Wilbur Schramm's model; Schramm 1955). Debray indeed goes beyond all these limitations of code, encoding, decoding, or decoupling of a transmission: mediology provides instead a more systemic and dynamic, a plastic and relational perspective on processes of transmission across the borders of individual cultures and specific eras. The constructed, *scarce signals* so often at the center of transmission theories are exchanged here for the *rich and plastic situation* of actual transmission (*cf.* Schulze 2018: 88–94).

This focus on the manifold and often in the process even multiplying material constitutents of transmission—a focus on all the buildings, mechanisms, and apparatuses as well as all the developers, operators, or lobbyists—stresses the material complexity and relationality of a transmission in action. For Debray and for any mediologist there is not just one sender, one receiver, one message, one route of transmission. In contrast, all of these elements are materially complexified, multiplied, and highly specified. Mediology is, in the end, never interested in the perfect or flawless transmission from a skillful sender to some passively waiting receiver; however, mediology is interested in the many conflicts, the distortions and disturbances on the way through a wild and rebellious collective of materials and actors. Transmission is something highly improbable for mediology. Any successful process of cultural transmission therefore is an extraordinary and instructive event. It is this improbable and instable character that makes Debray's concept of transmission especially interesting for research concerning sonic signals and their transmission.

Translating mediology

The approach of mediology provides a number of terms, figures of thought, and of approaches that converge surprisingly with current methodological issues in sound studies. Core elements of mediology transcend the current corpus of science and technology studies, of media theory, or of communication research—and it pursues the goal of developing and providing new and surprising concepts for all of these fields. These new terms make mediology in its materialist verve useful to apply in sound studies and in our research on sound design. It might even be that the insights of mediology are even easier to apply to sound studies than to beaten tracks of communication research or of science and technology studies. In this section I would therefore like to sketch out possible sonic translations of mediology's concepts and terms.

Mediology is, to start with, an approach that concentrates on one microscopic, narrowly defined activity of transmission. Unlike most approaches in communication research or media studies it is *not* analyzing how a single step in a mediation process takes place via signal processing, for instance. Instead mediology focuses on truly long-term transmissions in cultural history, *long-term* meaning here decades and centuries—however, such a "process of transmission necessarily includes acts of communication" (Debray 1997: 7). Mediology hence does not analyze how one sender tries to reach one receiver, yet it analyzes broader issues of transmitting fundamental ideas of a whole cultural context or a historical era into a still unknown future and a supposedly unfamiliar territorial area. This fact alone might seem a major obstacle in translating these concepts into a sort of *sonic mediology*. Though a second quality of mediological research makes it on the contrary very fitting for issues of sound cultures in general and sound design in particular. For in the center of Debray's mediology one finds a twin concept of how such a cultural transmission can take place: the double of *matière organisée* (*MO* or *organized matter*) and an *organisation materielle* (*OM* or *materialized organization*). The benefit of this double concept is that it is capable of adressing two perspectives onto processes of transmission that are routinely separated in research—not only into distinct methods but also into whole different, if not antagonistic, disciplines. These two research approaches are often split into *communication research* on the one hand and media history or *science and technology studies* on the other hand; but Debray's combinatoric play with letters (*MO—OM*) transmits here unmistakeably that both are complementarily intertwined. The first perspective on transmission analyzes the material substance, all the forms of how an *organized matter* (i.e., matière organisée, MO) provides the preconditions for transmissions; in this perspective the focus of analysis would lie, for example, on buildings and monuments, on data carrier materials, on new inventions in their material form, on newly invented tools to accelerate transmissions, or on transnational agreements on material standards in order for a message to be transmitted. The complementary and often separated perspective on transmission though analyzes its pragmatic substance, all the actions and procedures how a *materialized organization* (i.e., organisation materielle, OM) of actors manages to operate and to use these apparatuses and to transmit a certain cultural heritage into some unkown future and its unpredictable territorium. Apparatuses (MO) and actors (OM) are—from a mediological perspective—not actually separated ontological categories. Both are regarded as connected and intertwined servomechanisms contributing and subordinated to an overarching process of transmission.

Notorious examples in the field of mediology are therefore large institutions, frameworks, and infrastructures built to transmit cultural core concepts, values, ethics and pragmatics, epistemology and politics, even medical and philosophical operations and treatments. Debray's preferred examples are hence the whole of material edifices, libraries, monuments as well as the whole institutional, educational, judiciary, and also meritocratical apparatus of the *Catholic Church*, of *Marxist Orthodoxy*, or of the *Roman Empire* (*cf.* Debray 1981). In all these three cases the material means of transmission managed to reach future cultures and social contexts that exist centuries or even millennia from the establishing era of these cultural frameworks. Following Debray their transmission was being enabled exactly by a construction, maintenance, continuous reflection, adaptation, and even transformation of the foundational institutional structures; these structures include all the rituals and codifications as well as the physical buildings, artifacts, and apparatuses that document and confirm until today the impact of these three reference systems. Needless to say that in all these three cases the original writings or referenced political and philosophical concepts were *not* left unaltered. Yet, the basic goals, beliefs, and values seemed to be basically stable: law systems of later eras still rely on juridical models from the Roman Empire; a series of speech acts and idiomatic phrases of subsequent centuries rely basically on some already established by the Catholic Church or the Marxist Orthodoxy; transport routes of today follow still some of those of the Roman Empire; even everyday practices from the Marxist Orthodoxy or the Catholic Church are still in use, although altered. Mediologically speaking, in all these three cases of the Catholic Church, the Marxist Orthodoxy, or the Roman Empire a successful transmission was achieved.

But what kind of transmission is taking place when one is speaking about sound in general and sound design in particular? Or, in other words: *how may one describe the transmission of sounds in the multilayered perspectives of sound studies by terms of Debray's mediology?* And, more imporant: how may a *sonic mediology* help in analyzing the processes and the effects of functional sound design in everyday life?

From the beginning it is important to exclude a maybe all too obvious translation of mediology, namely the one into interpersonal signal transmission of sound events, maybe through air: you speak or sing a sound—I can hear it, I might respond. For the very specimen of interpersonal transmission simply does not qualify at all as a form of cultural transmission in Debray's sense. This transmission can hardly bridge a timespan longer than a fraction of a second, let alone centuries or even millennia. The area of translation for a mediology of sound design therefore needs to be much larger. It needs to transcend the interpersonal limits into an overarching matière organisée (MO) as well as an organisation materielle (OM). Regarding sounding and listening this refers to the *auditory dispositive* (Großmann 2008; Schulze 2016a) as a whole including *material* edifices, libraries, monuments, inventions, tools and machines as well as *institutional*, educational, judiciary, codificative, and meritocratical systems. A historical example for this kind of transmission of a whole sound or music culture can be found in the whole reference system of the bourgeois and European orchestra- and conservatoire-culture as it was established since the sixteenth century, at a climax in the eighteenth and nineteenth century and still prolific in the early twenty-first century. In such an analysis one would have to describe the construction and maintenance of specific concert halls and their training schools; the invention and standardization of certain seating arrangements;

of further developed instruments; up to the establishing of a powerful, global score industry; a market for scores; the educational, judiciary, and codificative structures—for example, regarding score writing, correcting, interpreting, selling—including the meritocratical system of prizes, titles, and functions in this system: from the *Kinderchor,* or Childrens' Choir (today maybe the *Musikalische Früherziehung* or Early Childhood Music Education) to achieving the title of a *Generalmusikdirektor,* the primary conductor and artistic leader of an orchestra, a choir, or a theater. This *conservatoire culture* pretty obviously achieved a rather stable transmission of their basic goals, beliefs, and values through its *material* edifices, libraries, monuments, inventions, tools and machines as well as its *institutional*, educational, judiciary, codificative, and meritocratical systems. But, moving away from this rather firmly rooted conservatoire culture into a still emerging and transforming *sound design culture*: are there similar mediological structures to be found in this field of contemporary sound design and sound production?

These structures in sound design as a profession can be detected first of all on the material side (matière organisée, MO) of the industrial development and distribution of new hard- and software needed to produce sounds. Connected to these material products there are various workshops, product introductions, bulletin boards, mailing lists, social media profiles, YouTube-channels, and tutorials to provide the knowledge for how to work with these new hard- and software products. These material devices for sound recording and sound producing apparently bring their own network of teachers, instructors, experts, gurus, demagogues, assistants and sherpas, missionaries, and solicitors already with them. This holds true also for various locations that need to be regarded as the material devices for professionally presenting, discussing, and performing the products of sound designs—be it at a conferences venue or in an open-plan office; be it on a designer's personal mobile phone, tablet, or laptop; or be it on a live-streaming platform, a more traditional concert stage, a sound art venue, or at a club. These materially specific locations, the edificies and technological infrastructures are more and more provided for and adapted to the needs of sound designers: at least punctually and on demand as we could note in various presentation situations of the sound designers whose work we followed. The structures on the side of the organisation materielle (OM), the network of people, though, seems to be harder to detect and to describe in the area of sound production and sound design in particular. Yet, here one can also find the aforementioned educational functions (YouTube-channels), also some judiciary practices (these days foremost regarding copyright contracting or licensing), programming and coding routines, even fashionable *sonic design patterns*; and even the first glimpses of a meritocratical system can be spotted, represented by international academic associations (e.g., the International Community of Auditory Display (ICAD)) as well as in various prizes and awards (e.g., the "International Sound Awards" and the "Audio Branding Awards" from the international Audio Branding Academy (ABA) in Hamburg, Germany), or even conventional titles for sound designers and sound producers. This reference system of sound design and sound production is, apparently, not yet as established and as refined as the system of conservatoire culture—it is after all still in its nascent state. In our fieldwork at various sound design departments and individual freelance designers' studios we could observe how these efforts to stabilize the new cultural reference system catch on. In this decade of the 2010s a certain rising recognition could be noted, which brought sound design from a rather fringy design activity to a more and more recognized one among the usual canon of design disciplines; the rise of the discipline of sound studies surely did not harm this development (Supper 2012). All of these efforts are precisely following the categories proposed by Debray, as they negotiate and establish larger associations, recurring meetings and proceedings, commonly shared research questions and production problems, as they award prizes and engage in a shared discourse on quality, on integrity, and on relevance. A mediology of sound design with its material associations of people (OM) and assembled materials in apparatuses (MO) can now be more and more consistently described as we try to do on these pages. A cultural transmission can be detected and it can be analyzed in its efforts, its achievements, and its successful or aborted trials. The mediology of sound design is emerging, step by step.

Three mediological questions

Researching the process of a cultural transmission—for example, in the area of sound design—starts with a series of questions. As early as 1992 Debray proposed three questions in his famous treatise on visual anthropology *Vie et mort de l'image: une histoire du regard en Occident* (Debray 1992); he elaborated these questions to a larger extent in another famous text, the chapter "Le double corps du médium" (Debray 1997: 15–36), translated as "The Medium's Two Bodies" in *Transmitting Culture* (Debray 2000: 1–20). In this section I will try to explicate these three questions in order to translate them into the terms and concepts of a sonic mediology and a *cultural theory of sound design.*

1. *What are the material artifacts involved in this process of cultural transmission?* Mediology is conceptualized by Debray as a research field that operates historically and materialistically. To analyze *mediologically* a given reference system and its forms of cultural transmission means to lay out at first the material substance, the specific objects, artifacts, the concrete locations, tools, machines, and their required and documented actions by which these objects of transmission are operated by:

> No tradition has come without being an invention or recirculation of expressive marks and gestures. No movement of ideas has occurred that did not imply the corresponding movement of human bodies, whether pilgrims, merchants, settlers, soldiers and ambassadors. And no new dimension of subjectivity has formed without using new material objects (books or scrolls, hymns and emblems, insignia and monuments). (Debray 2000: 2)

The materiality of a cultural transmission—as a constellation of objects, people, actions, gestures, and marks—is historically and culturally specific. There is no transhistorical or transcultural *invariant* situation of transmission. Even if various models for signal transduction and for interpersonal communication (e.g., Shannon, Hall, Schramm) seem to aim at such an abstraction: from the viewpoint of a *materialist mediology* such models can only appear as ahistorical and acultural models, mainly documenting the cultural transmissions of their period (e.g., information theory, critical theory, mass communication studies). Recent research in science and technology studies with a strong materialist foundation confirms this critique—for instance in these strong and convincing words by Jonathan Sterne:

> claims about the transhistorical and transcultural character of the senses often derive their support from culturally and historically specific evidence—limited evidence at that. (Sterne 2003a: 18)

Analytically, such a focus on the historical and cultural specificity of the senses, of their idiosyncrasies even, all of this leads to a focus on concrete materials; at the same time this very focusing on idiosyncrasies and specificities of materials also fuels dialectically an urge to transcend exactly these historical and cultural limitations. It is therefore *exactly* the material and sensorial specificity and limitation of transmission processes that promotes an effort to transcend them as incredibly attractive:

> We transmit meanings so that the things we live, believe, and think do not perish with *us* (as opposed to *me*). (Debray 2000: 3; emphasis in the original)

The first of the three questions in mediological research leads consequentially to trace, locate, situate, describe, and try to understand in reseach how the material objects involved in one reference system for cultural transmission function and are generated—for instance in the field of sound design in general or in one aspect of sound design, for example, the sound design for an *automotive navigation system*. This means to describe the machines and instruments, archives and tools, the communication platforms by which sound design is developed, presented, archived, being criticized, granted awards, and implemented for a certain product. Transmission is not an accidens: "Legacies are never the effect of pure chance" (Debray 2000: 5). The material artifacts involved in a process of *sound cultural transmission* hence must be described and scrutinized in all necessary detail.

2. *What are the meanings realized by this process of cultural transmission?* This question is probably the strangest from the viewpoint of transmission as a process of signal transduction or of interpersonal, mediated communication. From these viewpoints transmission must be ignorant to meaning, it constitutes more a timeless, ahistorical, and cultural invariant apparatus of transfer that any known culture of humanoids needs to apply in the one form or the other. For Debray this idealistically abstract model of transmission is not only useless, as it is too vague and in its imprecision actually more wrong than right, such a model fails already with its aim to describe transmission *generally, transhistorically, and transculturally.* A mediological analysis can only start with the culturally and historically specific meaning as incorporated in the material substance of a process of transmission: the network of military transport roads of the Roman Empire; the good news of the life of Jesus and its teachings as transported into the third millennium by the writings and rituals of the Catholic Church; or the economical and sociological analysis as well as the steps of a revolutionary process as proclaimed in the writings and public manifestations of Marxist communism since the mid-nineteenth century. According to Debray the meanings of these three reference systems are not unrelated or in an arbitrary relation to the means of transmission. The specific materiality of cultural transmission is a result of the historically and culturally specific meaning that called apparently for a strong and longlasting form of transmission.

With this focus on specific meanings as transmitted by specific means of transmission also specific realities of their actors, their inventors, their operators, their users—or in older terminology their *senders and receivers*—are being connected and bridged. This main insight of mediology hence leads away from any attempt to disconnect strategies of *sending or receiving* from their actual goals, their interpretations of culture, their lifestyles, their politics, their main philosophical and pedagogical interests. Transmission in Debray's sense is never innocent. Cultural transmission cannot take place in isolation from any contents transmitted. To the contrary, *only* in a specific form of transport the contents, goals, meanings, or issues find their lasting and materially stable form of transmission.

This second question of mediological research would thus result in the tasks to trace, describe, and analyze the actual meanings and realities involved in this specific cultural transmission. In our example this means to describe the meanings attached to the sound culture around designing sound and crafting sound art or artifacts for audio media communication.

One would find certain lifestyles, certain goals and meanings either referring to experimental musics, to strands of soundscape composition, or sound studies, to acoustic ecologies or to an auditory culture of refined listening, maybe of audiophilia or of an educated listening, trained in soundwalking or soundscaping. Chapters in Part Two, "Sonic Labor," especially Chapters 6 and 8, "Skills and Habits" as well as "Portfolio and Presentations," provided some insights into the lifestyles and underlying meanings transported with sound design.

3. *What are the actors, their allies, and their strategic goals of a cultural transmission?* According to Debray a transmission is made and procured by and through material institutions that last and care for this *longue durée*. Due to his personal biography in the political struggles between communism and socialism (being a consultant of Che Guevara and Fidel Castro in the 1960s and of François Mitterand in the 1980s) it might be that Debray in his outlines of mediology is very aware of the conflicts, the fighting lines, strategic movements, alliances, battles, and ceasefires that constitute such a longer lasting process of transmission:

> This is what makes each transmission work itself out polemically, and it requires a strategic competency bent on striking up alliances, filtering out and excluding the extraneous, hierarchizing and co-opting, drawing boundaries, and so on. The process can be apprehended as a struggle for survival in the midst of a system of rival forces either canceling one another out by disqualification or swallowing each other up. (Debray 2000: 5)

Mediologically, any successful transmission is never just the outcome of an arbitrary constellation; it is the willfully and often vigorously promoted result of a collective and highly strategic effort to build, to institutionalize, to guide, to formalize, and to implement the material artifacts as well as the related pragmatics in order to transmit a certain meaning, a set of propositions and arguments into a closer or far away future. It is definitely not a task on a daily basis but a long-term engagement into building a larger archive, its rituals and exercises, confirmed by daily repetition and institutionalized in a larger collective. Or, in Debray's concise aphorism: "Journalists communicate, professors transmit" (Debray 2000: 6).

The last and third question constituting a mediological research process would therefore be to trace, to find, to describe, to discern, and to analyze the actually manifold actors—publicly dominant or camouflaged in remote distance—that contribute to maintain a process of cultural transmission. In our example this means to describe the publicly most visible speakers and writers on sound design, the inventors, musicians, and sound engineers, the people working for the administration and the execution of a sound design project, the key account managers and the project managers basically running the whole business culture around designing sound and crafting sound art or artifacts for audio media communication; in brief: all of the actors who secure that sound design is not just an ephemeral media design section but has a lasting and more and more publicly recognized effect.

Mediology thus as a discipline can contribute to understand how the field of sound design evolves with its actors, how it secures its place in contemporary society, economy, and politics and how it can manage to produce further artifacts with sonic signals and implement sonic patterns and sonic affordances of various kinds into everyday life: in your car, your kitchen, on your computer, your mobile device. In the following and last section of this chapter I will therefore try to apply this mediological analysis to one area of sound design. To follow the route of one sound cultural transmission through history and across cultures this example will be confined to one specific area where sound design is very present in everyday life. This area has already existed for several decades and it still proves to be vibrant and prolific. Due to the long-term development of this area, it is rooted in and connected to several other areas of technocultural development outside the particular field of sound design itself. Again, in order to understand this particular field of cultural practice it is necessary to put it into a wider context of its time and cultural history. Hence, to understand the mediology of sound design in the twenty-first century I will exemplify this with a *mediology of the automotive navigation system.*

The mediology of an automotive navigation system

The material artifacts involved in an *automotive navigation system* and its process of cultural transmission (*mediological question 1*) with special focus on its sound design might seem trivial at first. In the 2010s one will find in a contemporary, state-of-the-art car interior usually a smaller or larger screen, a loudspeaker or a connection to the car audio system, as well as certain sockets to mount the system on the front console or the inside of the windscreen. However, this obviously constitutes only a small, materially tangible part of the whole system. The larger part of the materials necessary for this technique to operate is concealed and implemented in a cased hardware as well as on remote server farms. These remote artifacts need to be regarded as essential parts of the necessary material substance; they include also the archived scans and mappings of the physical roads and surfaces of the automotively accessible earth and the physically outlined formants of the sonic and vocal characteristics for voice assistants and sound signals. Contrary to common belief these characteristics and immensely large datasets are *not at all* immaterial: neither in their stored appearances nor in the way they are scanned and archived. This material presence begins with the unimaginably large number of cars scanning and driving the roads at any time for various global car manufacturers companies, navigation system producers, and streetmap providers; regarding sound design one needs to add the recorded voices, vocal characteristics, and auditory display databases to give drivers directions in a seemingly humanoid way. On top one must then add the various enterprises scanning and archiving stretches of land after stretches of land by the use of satellite systems as part of this globally expanded material substance of archiving and transmitting. All of these stored scans of sounds and streets, though, can only achieve a rather medium-term storage, in comparison to the long-term dimensions imagined by Debray. This long-term transmission, therefore, requires other forms of material artifacts to secure a sustainable transmission. This more sustainable cultural transmission of maps and sounds is represented in industry standards as provided by various globally active and influential manufacturers such as TomTom, Garmin, Clarion, or Magellan Navigation. A legitimate and convincing standard across all of these platforms apparently does not yet exist in the years we did our research—though companies like TomTom manage at least to provide tools to use their navigation systems in virtually all cars, and even in other transportation vehicles driving on roads and highways. The sound designs, voice characteristics and their underlying database system of vocal sounds, sentence composition, and conversation structure, though, might also in the near future still be regarded as part of the *company secret* and the *brand experience.* Only a cross-platform standard might pave the way toward a certain industry standard for the *auditory display* of navigation data—maybe connected to the already existing Navigation Data Standard (NDS). Only this standardization could then achieve a certain kind of cultural transmission regarding this sound design.

The meanings realized by the process of cultural transmission attached to an *automotive navigation system* with special focus on its sound design (*mediological question* 2) are mainly to be located, on the one hand, in all the assembled apparatuses (MO) for mapping and for voice assistants and, on the other hand, in the sensory and corporeal practices (OM) of navigating this new and contemporary practice of driving with an audioguide. To provide a navigation system of this kind for globalized and networked societies, then, includes two underlying, more general goals: a pervasive territorialization of this planet—as well as an evermore refined crafting of artificially generated voices. If one combines these goals now with the ubiquitous practice of assisted navigation and electronic route guidance systems in general (Warnes 2015) the main meaning emerges out of this distributed network: inbuilt into every navigation system is the anthropological concept of encasing everyday lives in a supporting structure of a vector mapped and quantified planet as well as being guided by an encyclopaedically informed voice assistant—that can relate at all times, in all situations of everyday life to this knowledge of the planet. This goal would qualify also as one of the main goals behind interaction design, database development, and visualizing and sonifying routes and landscapes. In all of these artifacts and in the way they were and are crafted the totalling and territorializing desire of control about every single square inch on this planet is apparent: a clear symptom of the *panacoustic society* that will be discussed in a later chapter in this book (Chapter 15: "The Panacoustic Society"). By its own and factual main apparatuses and activities the supposedly innocent tool of an automotive navigation system is, therefore, being redirected to the activities and apparatuses developed and applied in military operations and army dispositives. The goal to gain strategic positions on a battlefield—as part of a sequence of attacking and defending actions constituting a rather distinct series of battles that are again constituting a war—is quite ridiculously misappropriated to the everyday life activity wandering or driving around in a yet unkown urban or suburban landscape. The military dispositive built deeply into navigation systems recalls the military dispositive built into contemporary sonic signals (as explicated in Chapter 9: "Signifying Sounds," especially in the section "The semiotics of hunting signals"). However, should driving a car actually resemble using weapons in order to fight an adversarial army? The *sonic design patterns* that can be found here—being implemented in hardware, software, and performatively realized protocols by individual drivers—make a remediation and reactualization of military signaling and military pathfinding almost unavoidable. The material of cultural transmission is never meaningless: it also transports and promotes the earlier, maybe forgotten, camouflaged, or repressed meanings of these activities and apparatuses. The habits and skills, the conflicts and heuristics of earlier times are not actually erased, they are only hidden in concealed strategic goals.

These strategic goals and the main actors of this cultural transmission (*mediological question 3*) can hence be found in the main companies and corporations that are globally mapping the networks of roads as well as archiving, mining, and presenting selected excerpts with audiovisual guidelines for supposed driving actions. Starting from a simple, even boringly tiny tool of everyday usage this analysis is now entering one of the surely largest contemporary mediological artifacts. Corporate actors such as Google or Apple are indeed the major forces promoting a network of scanning, storing, distributing, and adapting the known world and its roads and streets to the necessities of a car. Obviously, the whole expansion of this into the product of a so-called *self-driving* or *autonomous car* then attaches this object of mediological analysis to an even larger and supposedly highly transformative commodity that might be entering our everyday lives much, much sooner than you or I might expect right now. As a consequence, it would not be too far fetched to regard indeed this very goal of achieving and socially implementing an uncountable swarm of self-driving cars as the actual and inherent strategic goal of an automotive navigation system: in the self-driving car the automotive navigation system comes into its own. The auditory display of this near future commodity is therefore not independent from these strategic goals, but subordinated under their main activities and apparatuses for mapping landscapes and storing vocal characteristics. The total territorialization of the planet and humanoid voices also applies here. The quite probable continuity between the automotive navigation systems of today and the self-driving cars of tomorrow might lead us therefore to the conclusion that the sonic pattern languages for voice assistants then will follow the already layed out path for sonic pattern languages for voice assistants today. Hence, one is safe to assume that the digital voice assistants as well as the sonic signal-sets implemented in mobile devices of today might end up in one, several, or even all of the sound designs for self-driving cars of the future. This is the world that might be created for and culturally transmitted into the all-too-near future by this technology and this sound design approach: however, if this world is desirable and if these strategic goals contribute to a society, an economy, a culture you or I would wish to live in right now—this can and will not be determined by mediology. Mediology describes, analyzes, and evaluates the promising, successful, or failing process of cultural transmission of one material cultural framework into a yet unkown future. Mediology does *not* decide upon the ethical value of what culture exactly is transmitted: it assesses the consistency of transmission—not the ethical quality of the transmitted material culture. This question, an additional and rather ethical mediological question if you will, requires other references and other analytical approaches in order to invent different scenarios, also deviant, imaginary mediologies. The ethical question, a fourth question added to mediology, would then probably read: what new material cultural framework of values, concepts, and ideas can be assessed as ethical from the then dominant perspective of a certain contemporary cultural framework—and how reliable, normative, and biased can such an assessment possibly be? Are there mediologies of humanoid desires transcending their historicity and culturality?

12

Transcultural Aurality

Decolonizing sound

Sonic stereotyping

Sounds occurring around us are routinely assigned to specific and often habitualized patterns, functions, or clichés. You and I perform this assessment and assignment habitually. This *aurality of the everyday* might serve to allow quicker responses, habitualized reactions, and fitting follow-ups; but this aurality might also prevent a deeper understanding of the actual sound events—if such habitualized interpretations and reactions prove actually to be very misleading and maximally distorting. Then the sonic pattern language used by a designer or a sonic prosumer needs to be revised or at least punctually corrected. Transcultural relations and migrations of sounds in particular seem to invite certain distortions and misinterpretations, and they are hence also calling for major revisions and corrections.

This problem becomes quite urgent if a sound branding is expected to represent multinational companies or a company's cultural diversity—and in the end it seems (as we could observe in several sound design projects) to be strongly drawn to cultural clichés. Take for example the sound branding we encountered in our research for a Europe-based company that constantly cooperates with experts from Asia. In a product presentation of this project we were presented a sound logo, two-seconds long, that is meant to represent the whole transnationally operating *culture of the brand*. But, to begin with, what did I *actually* hear? Listening to the sound logo, a somewhat blandly pleasing tiny melody reaches my ear, produced through a combined sound of a plucked string instrument, a calmly progressing piano melody, and some very faint rhythmic elements. I hear an assortment of clichés. There is an unspecific feel to the music, not annoying, not exciting. An association with some Asian tunes, some folk music maybe, but clearly directed at a white European audience. When listening repeatedly, the sound remains unobtrusive and sleek, there is nothing that sticks to me, nothing of a sing-along catchiness—the perfect easy listening, *a sonically domesticated cultural commodity for the perfect consumer citizen.*

What this logo-presentation suggests, however, is something very different. The sound design is presented here as based on extensive research of Asian and middle-European musical traditions that have influenced the design of distinct sonic patterns. The string instrument is a traditional European *dulcimer*, which should also remind the listener of Asian musical traditions in which similar instruments have a long musical history. These highly heterogeneous musical and cultural strands are reduced to a musical core tonality based on a pentatonic four-note sequence aimed at creating an affective connection—a connection of two distinct cultures, incorporated in the so-called *brand sound*. Also, a number of other sonic signals were found in this project that represent by convention various characteristics that a European- or Northern Hemisphere-audience might associate with Asian culture as a whole—a clustering of cultures and traditions that emanates a highly questionable exoticism: the sound of serenely flowing water, a pervasive absence of any noisiness, an additional use of a distinctly interjected piano melody. The sonic effects that leak from the designed sound logo shall resonate with a foremost Eurocentric listening habit that consumes this cosmopolitanized form of Asianness: *not too alien, yet not too familiar*. Surprising enough to raise eyebrows and generate interest, but still comforting enough in order not to discourage or to unsettle potential citizen consumers and employed prosumers. This artifact represents a sort of miniaturized cultural commodity, neatly packed and ready to be consumed and detached from any need to unfold the many entangled histories and more unordered, messy, and transnational cultural (sound) practices that are *not* captured. Taking this logo as exemplary for a lot of different sound designs, the task of sound designers would then seem to be to erase all stubborn, diverse, and erratic sounds in order to create a comprehensible, recognizable, and, above all, marketable core tonality: an interpretation that exoticizes sound design as this sound design does exoticise an apparently unfamiliar culture.

This fundamental essentialization of cultures as a whole still follows and stabilizes theories of *cultural circles* or *Kulturkreise* from late nineteenth- and early twentieth-century anthropology or *Völkerkunde* (e.g., Leo Frobenius, Wilhelm Schmidt, or Fritz Graebner) that contemporary research on cultural anthropology decisively rejects. This historical approach to cultural differences is regarded by the contemporary state of research as an aberration, mainly driven by imperialist and suprematist urges in the European nation states of that period. If applied though by sound designers of today this handy simplification can lead to the concept of a so-called *core tonality*, a term that is pervasive in the accounts of sound design practitioners (e.g., Jackson 2004; Bronner 2004; Bronner and Hirt 2009; Treasure 2010; Groves 2011; Weiss 2015). This term brings in the wish to essentialize a sound or a tone as intrinsically connected to one of the (again: historical and falsified concept of) Kulturkreise, completely ignoring the actual state of transcultural mingledness and interpenetration that characterizes the globalized and networked cultures of our times. It takes an almost unthinkable degree of ignorance and confinement to one's own everyday practices to speak of the—here I apply some of the clichés counterintuitively—genuine *European* symbolism of life and growth, the social and musical sound practices of a *European* farmers market or the deep spiritual qualities of a *European* monastery and its religious practices, as it would be to speak of a genuine *Asian* symbolism of life and growth, the social and musical sound practices of an *Asian* farmers market or the deep spiritual qualities of an *Asian* monastery and its religious practices. Sadly, these stereotypes will continue to have significance as long as people build their everyday racism on them—a stereotyping based on racist conceptions and the associated social, economic, and cultural oppression of people in many parts of the world that still impacts on racial prejudice and unequal distribution of power today. But, actually, they are largely void of knowledge or even empathy. Maybe some of the citizen consumers in Bangkok know more about your favorite European filmmaker and maybe you, the reader, practice a spiritually deeper concept of Buddhism than is actually common in the culture where one assumes both these cultural commons stem from? Actually, you perform, you actualize, and you revitalize these *cultural resources* (Jullien 2016)—if you live in Portland or in Thailand.

But such an essentialization and *stereotyping* is not limited to ethnic markers. Similarly, sonic patterns are created across other identity categories as well. This is for instance the case—focusing now on a different sound branding project of our fieldwork—when a linear progression is constructed in a sound design from the brand's fundamental denominators, as extracted from a company's visual and verbal self-descriptions, to the sound logo and sonic patterns of a brand. This conceptual and linear progression is the perfect example of a logocentric extrapolation: the claims and self-descriptions of a company, developed in long meetings of communications representatives, are to be taken literally and are then translated literally in sounds, in images, in videos. This *logocentric overreach* results in an ideological shortcut—or, as it is often legitimized in presentations, in so-called "evident and guiding characteristics." This overreach produces homogenous generalizations that are willfully blind and ignorant to the inner contradictions, paradoxical conflicts, and self-doubts in any cultural area. This results for instance in the common notion that anything somewhat related to *motherhood* immediately needs to imply *well-being* (though there are also many instances of diseases, disruption, and dread in motherhood), which then leads to *tenderness and generosity* (though there is also a large area of experiencing strength and vigor as well as a meticulous austerity in motherhood) which must be represented by a *female voice* that transports tenderness, confidence, and happiness (though female and motherly voices can also be extremely pervasive and harsh, and there are male voices and third gender voice that might transport tenderness in different and possibly more complex way), and the outcome is then a vocal performance with the mentioned sonic patterns that seems to be able to sell *well-being* (which is only to be found, apparently, in the form of pharmaceutics—which are in contrast to the supposedly genuine *well-being of motherhood*, at the starting point of this not-so-logical succession, right?). This sequence of concepts that are claimed to follow logically from each other are characteristic for the logocentric arguments (or, more precisely, the *simplistic and fake wannabe-logic*) in design presentations in all areas of the so-called creative industries and in the consulting business as well: a quite obvious series of horribly distorted conceptual round-off errors that pile up on each other. *Sonic stereotyping* is our name for this logocentric round-off error in the field of sound design.

There is a striking imbalance between the promise of the magic trick in sound design (*cf.* Chapter 1) and the blatant reductionism bound to a brand's self-promotion for sonic consumers. Sonic pattern languages that emerge from such processes can never actually be "culturally specific" sound designs. They present stereotypical sonic patterns indulging in an imaginary, more self-complacent sonic essence projected onto a certain group of humanoid aliens and their lives. The effect needs, apparently, to be just big enough to be catchy, recognizable, repeatable. Cultural and racial stereotypes are, hence, reconstructed and fixed, closed and confirmed, blown-up and amplified to the largest extent. The sonic pattern language emerging from such design processes refers then usually to common business practices and the notorious *will of the consumer citizen*—who obviously is very different from all the team members present, they would never follow such a reduction, but the consumers surely will, they can't help it. Reduction, repeatability, simplicity, and a core tonality is proposed to consumer citizens as the only form of communicating: a form of communication that I would call *strategic essentialism*. This essentialism is not to be believed, not to be lived, not to be embodied by the designers, creative planners, or developers who promote it; but it is used as a strategy in order to convince, to seduce, to overwhelm the sonic consumer. It is a strategically effective sonic pattern that intends to avoid or at least to mainly exclude all unwelcome forms of performing and appropriating this new sonic pattern. Strategic essentialism tries to anticipate and to abort the necessary sonic domestication by the citizen consumer. Sonic stereotyping therefore intends from the start to guide and to detour the sensory appetite of the sonic consumer (as explicated in Chapter 1, section "Functionality as guidance") *to digest, to embody, and to finally appropriate the expectation and the adequate reaction to these sounds.* Not the joy of a sonic persona or a lust for excess is fed here. But a sonic pattern language of this kind represents not much more than a conceptual and logocentric bricolage of cultural stereotypes. It is actually one of the reasons that sonic signs more often than not induce mainly a kind of boredom and at best a friendly ignorance in their listeners and users. They lack sonic joy.

Such a design practice repeats and stabilizes mainly the learned, outdated and horribly insulting, discriminatory, condescending, and debasing concepts of *"The Other"* that European- or Northern Hemisphere-traditions have accepted to regard as *real,* as *felt,* as *obvious*, or *self-evident*: the common markers of a stunning lack of self-reflection. Sonic stereotyping affirms and stabilizes these allegedly wrong and logically as well as socially inacceptable notions. As a sonic consumer I learn, hence, to read such sonic signals as part of a symbolic order I know is horribly wrong. This order can be as detached from my personal or individual experience as possible, it still represents the logocentric matrix of *"Truth"*: a strum of a guitar means *Western rock music*, a dhol drum pattern means *Indian folk music*, a tenderly metallic and sensibly self-centered voice with harsh electronic sounds represent *Nordic songwriting*, and a tender plucking of a Koto, a zither, represents *Japanese traditional music.* This stereotyping of a pattern language is therefore also taking place in the transposition of sonic signals of military music and hunting signals to the realm of everyday life, entertaining media events, and popular culture. These sonic affordances are deeply colonized by the historically imperialist concepts of earlier decades and centuries. Yet, how could it be possible to decolonize these sonic patterns and their sound design processes? Is it even thinkable to work in sound design for a sound culture and an aurality that does not indulge in stereotyping but in transculturally understandable concepts of sounds and listening?

Decolonizing sound design

The history of sounds and their design is deeply attached to the political and social transformations of the societies they take place in. Global westernization and the recently more and more questioned hegemony of Anglo-centered cultures is no exception here. An intrinsic aporia characterizes this process of globalization when standardization is, on the one side, needed to ground a wide-ranging, intercultural exchange but, on the other side, is blamed for a reductionism and simplifcation that is as wide-ranging, even destructive of cultural heritage as well. But does one *really* need such a strategic *sonic stereotyping*, relying on essentialisms in order to continue this ongoing exchange? Or isn't this more of an unquestioned truism, tediously repeated, reiterated—but never actually proven? More recently designers and artists also recognized this unsettling aporia:

> Mainstream design discourse on global platforms has always been dominated by a focus on Anglocentric/Eurocentric practices and ways of knowing ... there has been little attention paid to alternative and marginalized discourses (around, for example, gender, race, culture, the production of knowledge, material practices etc.). (Ansari et al. 2016)

With these words the research collective of the *Decolonizing Design*-project describes the controversial and critical situation of design in the early twenty-first century. The problems touched upon here arise from recent developments in design practices and design professions being tightly chained to industrial production and major corporations as well as their mannerisms and established standards of production. In the logic of industrial production, the process of stereotyping factually takes up a major role. A critique to *decolonize design* is hence timely and direly needed. Yet, aside from a critical practice constituting new sound design practices, it is our goal in this section to explore and to analyze some approaches toward *decolonizing sound design*. Such an analysis should provide further inspiration for sound designers and sound consultants of how to steer a sound design process, how to provide an environment for sonic labor that leaves false stereotypes and truist essentialisms behind. So, how to decolonize sound design?

Together with you, dear reader, we would like to now listen to a performance by Mendi + Keith Obadike with the title *Numbers Station [Furtive Movements]* from 2015 (Obadike and Obadike 2015). As soon as we enter the performance—physically or in an audiovisual recording—we hear the artists alternately reading a series of somewhat abstract numbers: "009, 383, 010, 277." Right now, these numbers do not seem to make any sense to us. But still, being dedicated listeners, we continue to listen: "013, 167, 014, 409." The sequence of numbers creates patterns. They raise doubt and irritation in us; we continue to listen. We wish to understand. "017, 92, 018, 167." Underscoring the reading is a composition of layered sine tones of different frequencies fading in and out. "019, 105, 020, 157." The repetitive uttering of the numbers by alternating a female and a male voice, soft and clear, together with the sustained sinus tones and feedback sounds create a sonic pattern that is both enchanting and unsettling. "023, 379, 024, 175." On the one hand, I want to draw closer, listen even more attentively, get submerged in the affective space this performance seems to offer. On the other hand, it makes me confused to try and understand the numbers that stay fragmentary signs, almost stubbornly resisting to make sense. "025, 255, 026, 212." The narrative remains incomplete, and I feel lost. And mixed into this feeling of being lost, while the numbers and rhythms and vibrations and voices go on, is an awareness that it is exactly this absence—of knowledge, of histories, of what *really* happened—that is the crucial point here. "028, 190, 030, 186." And it is only now that I can relate what was sonically, sensorially, and materially incorporated in the performance to the information that was offered to me about this project: the self-reported Stop and Frisk data of the New York Police Department from over 123 precincts is sonified, Mendi + Keith Obadike read aloud the number of the quarter and the number of stop requests alternately. "032, 048, 033, 452." The underscoring sound is generated from this same NYPD database, taking numbers of stops and translating them into frequencies. This installation and performance is part of a series of alternative *data mining projects*: data mining with goals different from the wish to document, to territorialize, and to control every consumer citizen—but with a desire to make audible the questionable if not totalitarian activities of surveillance and control of the public sphere. "034, 530, 040, 424." Both artists are interested in the ways that sonification of these data reveal something about what we get to know—and, importantly, what we don't get to know—about the sensory, racist, violent, and unheard-of aspects of the stop requests. As adequately described in an online art magazine:

> There was no way to link each set of data points to any one person. However, knowing that an overwhelming majority of that data represents actual black and brown people was a frightful reminder that the truths about inequity cannot be hidden, even when they're masked by bureaucratic numbers. The Obadikes' staccato, monotone reading voices made it clear that the subject of the performance was indeed the data, not the artists themselves. (Lynne 2015)

The subject of this performance was the data. However, it was a sonic pattern that resonated underneath, crafted by a humanoid, in a sensitive and somehow loving artistic practice, so that the data could play out its capacity to disrupt and to disturb the well-known stereotypes, the toxic essentialism, and the racial violence of this police activity. "041, 274, 042, 176." Sonification of data is turned in this case into a *decolonizing strategy*. A functional sound is not just repurposed, refunctionalized, and thus resignified, but it is *through* the adding of sound and voice and context with a clearly functional goal that new layers of meaning of the data are revealed and become public: a matter to be discussed, to hear, to feel. The sonic presence emphasizes a semiotic absence. "043, 207, 044, 334."

The goal of *decolonializing listening* is one that has been very prevalent also in the artistic practice and theorizing of artist-activist-collective Ultra-red. One of the crucial points in their work is the way that they define, use, and refine their activism in relation to a number of listening practices, sound walks, sonic interventions, and listening sessions. In their book *Five Protocols for Organized Listening* from 2012 they describe how these listening practices, and the constant self-reflexive and critical mode, is constitutive for their activism:

> There is a saying among Ultra-red members that some times [*sic*] the fastest way to get somewhere is to slow down. This sentiment captures the dialectical tension between the crises against which we react and our actions that bring the status quo (even the status quo of activism) into crises. (Ultra-red 2014: 5)

This saying resonates with Obadike's work in a compelling and thought-provoking way, demonstrated in how both artists use a strategy of slowing down to thematize a very current and ongoing discriminatory practice. The repetitive uttering of the numbers of stop requests in a gallery space is surely *not* a direct action with an immediate impact. Instead, the discriminatory machinery is slowed down: it is being probed and investigated, scrutinized and reevaluated, even tasted, sonically, itself. This is a sensory practice, an example of sonic micro- or even nanopolitics (taken up later in this book, in Chapter 16, section "Nanopolitics of sonic labor"). Slowing down and focusing on one sonic nanofragment, one constitutent in these sonic pattern alters the whole relation of listeners or sonic personae in this situation of sound events and sonic traces. The sensory effect and the sensible ramifications of these sounds are thereby slightly tweaked and transformed. As soon as your listening refocuses, though, as soon as you listen differently in different contexts, you also alter the meaning of the sounds you are listening to. Maybe the control aspect of this designed sound is weakened? Maybe the *silencing dispositive* doesn't hold you that tight anymore and is no longer capable of suffocating one's urge to articulate a pressing issue (*cf*. Chapter 13)? The sensory dynamics of listening and sounding are not superficial qualities of sound: they are the most appropriate contact point for critique and change. Both works by Ultra-red and by Mendi + Keith Obadike operate in this area of *nanopolitics* (Hansen, Ploteghe‍r, and Zechner 2013; Schulze 2018: 163–166).

A third example that highlights the politics of sound design and the complex, critical and even uncontrollable aspects of listening practices is Pedro Oliveira's recent analysis of "the contingent sonic practices engendered by one particular jukebox, which belongs to a particular place and produces a specific set of sonic affordances" (Oliveira 2017: 106). In the center of his analysis is a specific jukebox standing outside a bar in *Lins*, a favela in the northern part of Rio de Janeiro. In the ethnographic work conducted in this area, it became clear that the jukebox does not simply fulfill its intended function of suiting the pleasure of some middle-class people who wanted to listen to mainstream music. Instead, this jukebox worked as a transmitter of *proibidão*, a subgenre of *funk carioca* with highly critical and provocative lyrics that speak up against the dominating military police—a fact that makes this music actually forbidden and causes this machine to be repeatedly confiscated. Moreover, the jukebox works as an important indicator of "the general mood, and more importantly, whether it would be safe to go out and walk around"—as long as the jukebox was on and playing, things seemed to be alright, but as soon as it went silent, this potentially meant trouble (Oliveira 2017: 113–114):

> The cultural and political function of this type of jukebox, and specifically the jukebox from *Lins*, subvert and extend well beyond its intended design, embracing the contingency of certain localized listening practices in order to become, in itself, an ambiguous device for provisional forms of auditory governances. (Oliveira 2017: 106)

The crucial point here in unfolding the designed object and detaching it from its "intended design," is the realization that the listening practices that develop around such an object have a life of their own—which means: as designers we create sounds for objects (or devices, or spaces, or situations) that might seem very suitable, compelling, marketable, or educational—but as soon as the sounds are out there we are no longer in charge of the effects, uses, misuses, and transformations that they set in motion. Therefore, Oliveira proposes also a *human jukebox* that expands this volatile and dynamic character and makes it an actual tool for resistance and for altering social contexts with a political impact.

With regard to the transcultural semiotics of sound design processes, such diverse cultures—as just unfolded in the examples by Ultra-red, Mendi + Keith Obadike, and Pedro Oliveira—and the roles sound plays in them become prevalent, both on the level of perception and on the level of production. In order to make the idea of cultural semiotics productive for the understanding of listening situations as well as for concrete sound design production scenarios, the focus therefore needs to be shifted from one rather atomistic and causalistic core assumption of semiotics: that one primarily reacts on, and responds to sounds in a semiotic process with foremost rather stable and unaltered constitutents. A productive application of cultural semiotics in this context would acknowledge a performative situation in which a number of heterogenous sonic signs come into play in one moment that are used generatively in crafting new sounds, used in establishing new sonic patterns, and—consequently—also at least situated interpretations of new sonic cultures.

The remediology of sonic labor

The three examples explored in the previous section share various approaches and techniques with which they manage to decolonize sound design. In this section we would like to collect some of these tactics that might provide a toolbox for designers and that might even lead into a reflexive decolonization of sound design in general. The three examples by Mendi + Keith Obadike, Pedro Oliveira, and Ultra-red can be differentiated in the following ways: (1) *shifting the sonic material* from a commodified and formatted to a situated and corporeal one (in the example by Mendi + Keith Obadike), (2) *shifting the situation of sonic labor* from a commodified and formatted to a situated and corporeal one (in the example by Pedro Oliveira), (3) *shifting the listening practice* from a commodified and formatted to a situated and corporeal one (in the example by Ultra-red). These three practices of shifting sound design and listening practices, situations of sonic labor, and the sonic materials result, apparently, in thoroughly different sonic artifacts. All three forms of decolonizing sound design shift the whole of sonic labor in significant constituents.

When *shifting the sonic material (1)* from a commodified and formatted to a situated and corporeal one, the sonic prosumer does leave predetermined limitations, pro- and inhibitions as well as clichés widely behind. Whereas one of the tasks of a sound design could be to design with a functional sound the sonic signals of the communication devices used by the NYPD, this is significantly shifted: the material is no longer only the functional and commodified sound, but the material here is suddenly the whole situation of communication in its situated and corporeal aspects with a multitude of social and political ramifications. In the example by Mendi + Keith Obadike the sound of police communication is not simply reorganized and reworked in a way that follows an ideology of efficiency (*cf.* Chapter 3, section "Ideologies of efficiency") and a hegemonic dispositive of silencing disturbing voices and noises (*cf.* Chapter 13); on the contrary, this form of sound design expands its reflected material to the whole situation of communication: Mendi + Keith Obadike actually work sonically with the whole sonic material that is to be found in the very situation and its corporeality, as transmitted by the actual technological tools. Whereas a (neo-)colonialist sound design would have simply followed the preexisting constellations of controlling sound (*cf.* Chapter 13, section "Controlling sound") and of militarily connecting consumer citizens (*cf.* Chapter 1, section "The sonic consumer"; and Chapter 2), along the lines of Karbusický's categories such as *orders, contact signals, informative signals, mood signals*, and *signals with heightened symbolic meaning* (Karbusický 1986: *passim*; translated by Carla J. Maier), this present form of decolonizing sound design documents, questions, and transfers these categories. They are not taken for granted in their essentialist immediacy, they are not accepted as an unquestionable reality, but they are transformed and transferred into a different framework. Translated into the terminology of Debray: the cultural transmission of the policing dispositive has been put to a halt as the material artifacts are questioned. Or, with the first mediological question (*cf.* Chapter 11, section "Three mediological questions"): *what are the material artifacts involved in this process of cultural transmission—of the police dispositive?* By the artistic intervention of Medi + Keith Obadike the cultural transmission of the policing dispositive has been at least punctually *remediated.*

In *shifting the situation of sonic labor (2)* from a commodified and formatted to a situated and corporeal one a protagonist in the sonic workforce alters the routines and standards of predetermined limitations, pro- and inhibitions as well as clichés of working on sound designs. Whereas one of the tasks of a sound designer could be to design a jukebox in the most smooth, functional, and controlled way, this is significantly shifted here: the situation of sonic labor is not focused on a mainly functional and commodified sound gadget, but it focuses now on the whole situation of sonic labor in its situated and corporeal aspects with a multitude of social and political ramifications. In the example by Pedro Oliveira the jukebox is not simply reinvented in a way that it follows an ideology of efficiency and a hegemonic dispositive of silencing any disturbing noises; on the contrary, this process of designing sound reflects the whole situation of sonic labor: Oliveira actually proposes to work differently in the sonic workforce by including the very situation and its corporeality. Whereas a (neo-)colonialist sound design of the jukebox would have simply followed the preexisting constellations of controlling sound and of serving consumer citizens along the lines of the forces of authoritarianism (*cf.* Chapter 15, section "Forces of authoritarianism")—*consumption, dispositive, domestication*, and *ubiquity*—this present form of decolonizing sound design questions, alters, and reinvents these categories. They are not taken for granted in their assumed innocence as mediating function, they are not accepted as an unquestionable reality—but they are transformed and transferred into a political framework of decolonization. Translated into Debray's concepts: the cultural transmission of the jukebox has been put to a halt as the meanings realized with it are questioned. Or, with the second mediological question (*cf.* Chapter 11, section "Three mediological questions"): *what are the meanings realized by this process of cultural transmission—of the jukebox-device?* The meaning of jukebox-listening—as observed and analyzed by Pedro Oliveira—has been politically *remediated.*

Finally, *shifting the listening practice (3)* from a commodified and formatted to a situated and corporeal one a sonic consumer is altering her or his listening practices: moving away from the learned standards of predetermined limitations, pro- and inhibitions as well as clichés of listening habits. Whereas one of the tasks of a sound designer could be to design listening along the lines of its functionality of consumption (*cf.* Chapter 13, section "Functional listening"), this also is significantly shifted here: listening practices are not focused on radically functional and commodified practices, but they are refocused to its situated and corporeal aspects with a multitude of social and political ramifications. In the example of Ultra-red, listening is not simply reconceptualized in a way that follows an ideology of efficiency and a hegemonic dispositive of silencing; on the contrary, these newly proposed practices for listening reflect the situation of and the sensory corpora of their protagonists: Ultra-red propose to listen differently in everyday life including the very situation and its corporeality. Whereas a (neo-) colonialist sound design of listening would have simply followed the preexisting constellations of sonic consumers listening to controlled sound, this present form of decolonizing listening questions, alters, and reinvents these categories. They are not taken for granted in their goals, their protagonists and allies, they are not accepted as an unquestionable reality, but they are transformed and transferred into another framework. Again, translated into the concepts of Régis Debray: the cultural transmission of established listening practices has been put to a halt as their actors, their allies and their goals are thoroughly questioned. Or, with the third mediological question (*cf.* Chapter 11, section "Three mediological questions"): *what are the actors and their allied and strategic goals of everyday listening practices?* The actors and the goals of these listening practices have been—in this exemplary intervention by Ultra-red—*remediated.*

These three decolonizing shifts by means of mediological questions move sound design from a commodity supporting the aforementioned four *forces of authoritarianism—consumption, dispositive, domestication*, and *ubiquity* (Chapter 15, section "Forces of authoritarianism") into a new forcefield yet to be specified. These three exemplary shifts obviously are not attached to the development of a specific sonic pattern language or a general critique of stereotyped sound. But they are attached to the fundamental selection of sound design practices, labor situations, and sonic materials: as such the *resulting* sonic patterns or sonic stereotypes are hence affected and altered as well. Instead of proposing specific changes to specific sound designs, *the whole working process of designing sound is decolonized and refocused.* This refocusing, reorganizing, and reconceptualizing of sound design resembles in part the larger and intensely discussed process of *remediation* as explicated by David J. Bolter and Richard Grusin in 2000. Both authors explicate the practices of remediation on occasion of the coexistence of various media practices, with the example of a *webcam jukebox*: "our two seemingly contradictory logics not only coexist in digital media today but are mutually dependent" (Bolter and Grusin 2000: 6).

Translated into our issues here regarding decolonizing sound design practices, the *coexisting contradictory logics* are now not different generations of media but different forms of sonic performativity. In our case the surely quite *contradictory logics* of sonic signals in police communication or in communications about racial violence in police activity (Mendi + Keith Obadike), of a jukebox as music sales instrument or a jukebox as an agent of resistance, "a protester against the eventual gentrification and real estate speculations" (Oliveira 2017: 142–144), and of listening as functional or listening as explorative (Ultra-red) are actually *mutually interdependent.* Both logics can be developed out of each other—even as the latter, the political and situated ones stressed in this chapter, will surely lead some readers to a maybe disappointed question like: "But that's not *really* sound design, right?" Actually: *it is.* However, it is indeed *not a conventionalized form of sound design* for a supply industry serving institutions of controlling sound as well as of the silencing dispositive that addresses us as sonic consumers. This new and unconventional specimen of sound design is in contrast not serving a supply industry at all: *decolonized sound design* in this sense represents an actual form of sonic performativity that does not confine itself to provide sonic patterns or stereotyped sounds for requested commodities. It retains and it cherishes its ambition to actually contribute to societal aspects and corporeal as well as situated effects of sound. This *remediation of sound design practices, sonic artifacts, and sonic patterns* alters, hence, also the mediological process of cultural transmission. What has been analyzed, for example, on occasion of the automotive navigation system at the end of the previous chapter (Chapter 11, section "The mediology of an automotive navigation system") as genuine properties and liabilities of contemporary sound design is therefore not any longer applicable. In this earlier example we could detect various processes of standardizing mapping and guidance, also various sonic design patterns related to Karbusický's categories of the hunting signal and of voice assistants, and the overall project of global mapping and the project of autonomous agency in the self-driving car; in these three new examples here all of these earlier connections, dependencies, and linkages, all these liabilities, actually no longer apply, they are *delinked* (Amin 1985; Mignolo 2007; Ismaiel-Wendt 2011, 2013). Other agencies, other connections and linkages, intentions and agencies become meaningful and guiding instead.

The techniques and practices of decolonizing in these three cases result in a critical practice that reassigns a wide, completely new, variety of aspects, constituents, and practices. The whole production nexus of sound design, the sonic labor of a sonic workforce gets disassembled and reassembled in a different way. This process of disassembling, reassemling, of remediating sound design according to the lines of decolonization alters its mediological content of cultural transmission: this alteration I like to call a *remediology.* A remediology as critical practice results in a *reweaving* of its practice field—from scratch: from *delinking* to reweaving in a newly reflected cultivation.

Toward a transcultural aurality

Transcultural sound inhabits this globe. In the reality of the twenty-first century it is hard to imagine a strictly separated multitude of cultures—though there are, obviously, many sub-, micro-, or nanocultures that retain a set of cultural practices as their core that seems to show a certain, smaller or larger, gap, a discrepancy to other practices (*cf.* the term "écart" in Jullien 2016). However, the interweaving and interconnecting of global cultures is an everyday practice. According to this overarching transcultural condition it becomes necessary to address the issue of a variety of *auralities* and their transformations due to the decolonization of cultures as a whole, affecting all areas of life up to their sets of listening and sounding practices linked to their sets of sounding and listening apparatuses. Until now, though, the concept of *aurality* has not yet been used in this book. But at this stage of our discussion it becomes urgent to address its impact in relation to processes of decolonizing sound design.

The concept of aurality is different from others we have used up until now in this study. It is not completely identical with—as insinuated in the previous paragraph—*sets of listening and sounding practices linked to sets of sounding and listening apparatuses.* In contrast, this represents more a definition of a sound culture. Aurality, however, has a broader and thoroughly more abstract scope. The aurality of a historical period or a cultural context (e.g., Erlmann 2010 or Ochoa Gautier 2014) implies and defines not directly specific and material cultural practices and their apparatuses; aurality, in contrast, is more of a prerequisite for the whole of a culture that does not only affect its sound culture, but all aspects of the economy, of administration and governance, of finance and investment, of the arts and design, of the sciences and the humanities, of crafts and housekeeping, of everyday life. In the more narrow sense, aurality, refers to "the shared hearing of written texts" (Coleman 2007) as the most common document of historical and cultural knowledge. So, whereas a *sound culture* is to be excavated of the actual practices, the material culture, and the apparatuses dominant in a culture or a historical period, the *aurality* of a culture or a historical period represents a pervasive, underlying, and structuring force that might not even result in actual sound practices or listening apparatuses. The aurality of a culture refers to its main assumptions, its knowledge, and its ontological, epistemological, and anthropological insights and positions regarding listening and sounding—also confirmed by and discussed on occasion of actual sound practices and listening experiences.

A *transcultural aurality* now can only be conceptualized on the grounds of a decolonizing transcultural continuum—which, frankly, is more a desideratum these days than an actual reality. It is a goal to reach, not a constellation to start from. The goal of decolonizing aurality, hence, is as ambitious as the general goal of decolonizing this whole planet. It is a task and a project that is impossible to fulfill in a smaller, countable number of years or even decades: one can safely assume that the process of decolonization might be a largely infinite project - that will stay with future cultures on (or outside of) earth as an ongoing task and a responsibility. The reason for this lies in the sheer complexity, density, and depth with which colonization and territorialization are rooted in contemporary culture, politics, lifestyles, and economy. At least four major areas can be detected in which colonization is continuously present—and hence these four areas are the main objects for a process of decolonization:

1 The appropriation of land and the exploitation of labor.
2 The control of authority (viceroyalty, colonial states, military structures).
3 The control of gender and sexuality (the Christian family, gender and sexual values and conduct).
4 The control of subjectivity (the Christian faith, secular idea of subject and citizen) and knowledge (the principles of Theology structuring all forms of knowledge encompasses in the trivium and the quadrivium; secular philosophy and concept of Reason structuring the human and natural sciences and the practical knowledge of professional schools; e.g., Law and Medicine, in Kant's contest of the faculties). (Mignolo 2007: 478)

Is it too pessimistic to assume that a strictly martial and predatory, thoroughly *terraforming* process of colonialism as a redistribution of wealth and an annihilation of existing social structures takes at least as long to disassemble (if desired and pursued at all by former colonial superpowers) as it took to reach its most horrible and atrocious, warmongering climax? Regarding modern colonialism, incipient in the sixteenth century, one should thus reasonably not expect that a process of decolonization is one with a finite end-date. We are now situated in the midst of a continuous project of decolonization that affects all areas of work—and that will have to continue for the time being, maybe until the end of written human and nonhuman history. Moreover, the transformation of an *aurality* is an equally complex process; and in this case this is hardly to be done by executive orders or new legal frameworks. It is a process of cultural change that involves actually a double process of cultural transmission as discussed earlier (Chapter 11). This double, mediological process affects, as just mentioned, on the one side, the disassembling of the cultural transmissions of colonialism and at the same time, on the other side, an establishing of a new cultural transmission of postcolonialism, postracism of a more or less emancipated and equal exchange among nations and cultures, regardless of whether they are from the Northern or the Southern Hemisphere, skin color, belief system, or economic or military power. It is an impossible task, almost, for sure. So, what could be done to complete this impossible task, and what would stand at its end? What would be the characteristics of a transcultural aurality?

Decolonization as a process does affect sound cultures and auralities. The decolonization of listening practices and sounding apparatuses therefore needs to be an intrinsic part of this larger process. The aforementioned three techniques of decolonizing sound design can serve here as starting points. The three techniques of *(1) shifting the sonic material* from a commodified and formatted to a situated and corporeal one, *(2) shifting the situation of sonic labor* from a commodified and formatted to a situated and corporeal one, and *(3) shifting the listening practice* from a commodified and formatted to a situated and corporeal one, have obviously one double goal in common: a major change in all three categories of sonic material, sonic labor, and listening practices *from a commodified and formatted to a situated and corporeal one*. These shifts hence propose to leave established formats and dispositives, routines and skills, resources and archives behind—and open up to accept thoroughly new ones: new as they can only be found in the present and corporeal situation of listening and sounding—with differing centers and in new relations. These materials, forms of labor and practices cannot simply or willfully be selected, they need to be found, to be understood, and to be framed anew. This new framing however does not start from the established and formatted centers—but from experienced and present situations of wounds and borders, frictions and doubts:

> The geo- and bio-political shift in epistemology presupposes "border thinking" (which as a de-colonial project is always already "critical" but beyond Horkheimer's and post-modern uses of the term) and "border thinking" is the connector between the diversity of locals that were subjected as colonies of the modern empires (Spain, England, the US) or that as empires had to respond to Western expansion (e.g., China, Russia, the Ottoman Empire until 1922). Border thinking is grounded not in Greek thinkers but in the colonial wounds and imperial subordination and, as such, it should become the connector between the diversity of subaltern histories (colonial and imperial like Russia and the Ottoman empires) and corresponding subjectivities. (Mignolo 2007: 493)

This "de-colonial shift" (Mignolo 2007: *passim*) also implies that decolonizing techniques could be translated into various areas, fields of practice and areas of culture. In all of these new cases, again a return to the present and corporeal situation would be necessary in order to question established formats, clichés, ideological categorizations, and heteronomous relations of radical and involuntarily, mostly rejected, dependency and submission. It is also a process of desubmission and newly relating existing areas of culture to each other. This is what was meant earlier with the double term of *delinking and reweaving*: whereas *delinking* results in an "orientation toward pluri-versality as universal project leading toward a world in which many worlds will co-exist" (Mignolo 2007: 499), a practice of *reweaving* tries to start from this situation of delinked constituents and makes efforts to reweave, to rebuild these relations anew.

As soon as the colonized and territorialized constituents have been abolished, and maybe new ones found and the old ones reorganized, then the reweaving can start: a process of newly connecting, yet under completely transformed and dehierarchized conditions. It is by far though not a *nonhierarchical* situation—this must be regarded more of a conceptual and abstract imaginary, an imaginary point, useful for reference. Decolonial conditions are dehierarchical conditions as the previous hegemonic ideology has been dismantled, and there is now—maybe only for a brief moment in time and space—room and some time for detecting more appropriate, more situated and present hierarchies with a material reality.

A transcultural aurality that could stand at the end of such a process, is obviously not predefined in all parameters. The constituents of this decolonized aurality are however highly implicit in the process of reweaving the cultural interdependencies. Such a transcultural aurality, hence, cannot be centered in one hegemonic culture, dominant sound culture, or instructing aurality—though its protagonists will surely be rooted in one specific culture or aurality. With regard to the various aspects of "Living With Sound" we explored and discussed in this third part of the book, the outline for a transcultural aurality could be found along these lines as far as sound design is concerned: the process of "Signifying Sounds" (Chapter 9) could foremost be based on performing, rematerialized sonic signs and patterns in a process of "Situated Signaling" (Chapter 10) that generates meaning in situ and bound to the relational dependencies in order to contribute to "Transmitting Sounds" (Chapter 11) in the framework of a not labor-related, not control-dominated, and not silencing-euphoric mediology; with training in a framework of "Transcultural Aurality" sound designers might no longer be inclined to select or craft sonic stereotypes. This process of remediating and decolonizing could then be an ongoing practice and everyday task of long-term, foundational dimensions. Such a decolonizing design—in a broader sense of *Gestaltung*—would then take into account the whole cultural process of designing sound, including the working environment, the sonic pattern language, the uses and abuses of functional sounds in everyday life. Sound practices as activity and social praxis. This process of decolonization would then, necessarily, also affect research practices in respect to design: decolonizing research and decolonizing design research is again a process that is still to begin—also in our own future research.

PART FOUR

Sound Works:

A Cultural Theory of Sound Design

13

The Silencing Dispositive

Listening as function

Admission, presence, documentation

Sound design in the early twenty-first century is constituting and promoting a silencing dispositive of controlling sound and of functional listening that serves the pervasive surveillance processes of admission, presence, and documentation: this is the first major result from our fieldwork at sound design projects of differing scales and of varying applications as well as our critical analysis of the history, the design approaches, and the listening practices of sound design. The empire is designed and it is closed. Are there any locations or at least certain zones in everyday life where the various dispositives of contemporary globalized consumer culture are *not* the dominant and predetermining forces? These dispositives include dispositives of renting and buying, of selling and lending, of commuting and delivering, of billing and paying, of various transactions regarding material meso-objects as artifacts of everyday life and as material micro- or even nano-objects not visible and touchable in everyday life. These and many more categories of artifacts shape your life and mine, they even dominate to a certain extent my and also your activities, your professional life, and intimate life to a larger extent, too. They even must be regarded as the only and exclusive center and aim of your and my actions in almost every single minute of our lives as trained and orderly behaving consumer citizens. This is your life now. This is my life now.

Seemingly, we are not much more than processing units, fulfilling certain tasks regarding artifacts. But are we really? Is there truly nothing more? Is this bleak description not only the result of an uninventive and maybe depressed imagination, but also triggered by contemporary developments regarding pervasive surveillance activities and expanding authoritarian parties and governments on this globe? You still have your thoughts, your holidays, your garden, and your adventurous trips and journeys into maybe less commodified territories and regions, you still might enjoy excessive and sexually transgressive encounters, maybe induced and promoted by the intake and consumption of an assortment of recreational and performance-enhancing drugs, right? Or, as some approaches to understand these functions of contemporary societies of control seem to suggest (e.g., Lau 2000; Fisher 2009; Preciado 2013; Kinnamon 2016), are these supposedly last resorts of individual freedom or agency not some sort of granted as well as commodified and gated playgrounds, providing certain areas for release, for relaxation in order to provide more obedient and happily functioning consumer citizens? In any case, it seems indeed not too farfetched to speak of complex, interlocked, and inbuilt architectures of control: yet, not as a metaphor as it occurred recently on a regular basis in urbanism and critical theory; but as an actual and material description and a precise definition of contemporary infrastructures in cities, in buildings, in media apparatuses, in workplace environments, in environments for relaxation and recreation, and—last but not least—in systems for transportation and border control. This control is executed in three different functions, layers, and demands—that each relate to or appear mainly in sonic percepts:

(1) *Admission:* you, the consumer citizen, are required to plea for a *confirmation of admission* to a certain spatial area (on/offline) of commerce and consumption; you await, hence, the sonic signal of admission, an *informative signal* as well as a *signal with heightened symbolic meaning* (Karbusický 1986), crafted by a corporate design project.

(2) *Presence:* you, the consumer citizen, are required to *be physically present* in certain spatial areas (on/offline) of the administration of commerce and consumption in order to retain your status of being a consumer citizen with all its rights and duties; and again, you expect a sonic signal confirming your presence, surely a *contact signal,* but again probably a *signal with heightened symbolic meaning* (Karbusický 1986), crafted as well by the aforementioned corporate design project.

(3) *Documentation:* you, the consumer citizen, are *required to deliver documentation,* certain proofs or artifacts resulting from your activities to a previously assigned dispositive (on/offline) in a precisely prescribed and in each detail specified format; again you will hear either a sonic signal confirming or rejecting your documentation, an *informative signal* and also a *signal with heightened symbolic meaning* (Karbusický 1986), conceived and implemented by the same corporate design project.

These three main modes of control—*admission, presence, documentation*—and their sonic signals are built into the operation system of globalized, intensively mediatized and networked labor societies of the early twenty-first century: be it in public transport or in state administrations, in airports or in shopping malls, in gated toilet stall apparatuses or in amusement park entry systems, in refugee camps or in first class lounges. A society *not* executing these forms of control is simply not regarded as a functioning and properly organized society these days. Sonically, these three layers are escorted in ways even more distinct than just the semiotic descriptions attached. As a consumer citizen you receive (a) *the informative signal of your admission* largely in the form of a pinching, high-pitched sound, often mimicking or identical to a sound of a piezoelectric speaker or buzzer; (b) *a contact signal that documents your physical presence* largely in the form of a joyful sequence of tones, maybe to soften your tense insecure mood of awaiting this documentation; and (c) an *informative signal with heightened symbolic meaning* confirms or rejects your documentation by functional sounds suggesting an alarm, a failure, or a satisfying, machinist digestion following your timely and correctly formatted delivery.

Hence, the multilayered and processually documented process of control is asserted, executed, and communicated in continuous articulations of sounds or sonic particles as well as in various sensory signals (such as buzzers, etc.). This process as a whole constitutes a *dispositive* in the materialist sense as proposed by Giorgio Agamben, building upon Jean-Louis Baudry and going way beyond the more common reference in the work of Michel Foucault (Baudry 1970; Baudry and Williams 1974/5; Foucault 1980; Agamben 2009;

Schulze 2016a). In this definition, the notion of a dispositive is not restricted—as proposed by Michel Foucault—to travelling ideas and concepts, persisting figures of thought, or larger, overarching concepts enveloping longer historical periods and larger cultural areas. In the conversation *Le Jeu de Michel Foucault* (translated under the title: *The Confession of the Flesh*) Foucault stated:

> What I'm trying to pick out with this term is, firstly, a thoroughly heterogeneous ensemble consisting of discourses, institutions, architectural forms, regulatory decisions, laws, administrative measures, scientific statements, philosophical, moral and philanthropic propositions—in short, the said as much as the unsaid. (Foucault 1980: 194)

Apparently, Foucault expands the materialist, technological, and Freudian notion of the dispositive or apparatus—as established and discussed originally by Jean-Louis Baudry—into a notion suitable for his project of a new history of ideas. This Foucauldian interpretation sticks with the term dispositive still today. However, in the materialist interpretation that we propose here, a dispositive embodies such ideas and materializes them foremost in concrete, in massive, and in almost irrefutable structures, deeply anchored into the fabric and the foundation of a society, of a culture, of a political entity. This fundamental notion of the dispositive has had a strong impact on long stretched periods of cultural history: only slightly beginning with a history of recent decades, maybe a century, and not at all ending with histories that overarch several millennia, several downfalls and emergences of new political entities and larger cultural forms of governing, producing, trading, and everyday living. The impact of the dispositive was, in the end, the emergence of a commodified, massive product-centered society and culture. The things, the apparatuses, and the material infrastructures in which one lives in the twenty-first century tend to overshadow if not predetermine all possible, desireable, or inteded activitites. This concept of dispositive is also connected to Régis Debray's mediological concept of *organized material* (i.e., *matière organisée, MO*; as elaborated in Chapter 11, especially section "Translating mediology"). This foundation of cultural transmission, according to Régis Debray, includes buildings, data carrier materials, and all newly invented tools to accelerate transmissions. Architecture and infrastructure are main conditions for activities of humanoid aliens: without this very material dispositive, it would obviously be impossible even to write this book on sound design. Each living room, each flagship store, each larger area of a public square is organized according to predetermined structures that also bring with them certain predetermined forms of auditory or sonic interaction. Being in a living room, in a flagship store, or in a public square determines, hence, to a large extent the socially, culturally, and juridically acceptable activities one might engage in in these locations. In such dispositives, any sounding is assigned and required to execute a certain function as part of this dispositive, also any listening is assigned and required to execute a certain function as part of this dispositive: in the dispositive of cinema listening you might (almost) never sing or hum along to music being played—unless it is a required activity of this specific movie or music event. Obviously—as demonstrated in Chapters 5–9 of Part Two in this book—only the individual actors, their association, and their actual performativity actualizes, uses, and animates these rather dead apparatuses (this constitutes the *material organization*, i.e,. organisation materielle OM, according to Debray). However, the grounding quality of the dispositive is undeniable: one can actually bypass, adjust, or reorganize a structural dispositive over time in order to adapt to new contemporary cultural developments; but one can hardly consistently and enduringly perform completely outside of any apparatus or dispositive. In such a governing dispositive any practice of listening or sounding that serves foremost an aesthetic or intrinsic hedonistic purpose is easily regarded as irrelevant, disturbing, dangerous, superfluous, or useless. Try joyfully singing in a border control queue; try whistling while under surveillance in a tense and neatly organized situation of public transport; try car alarm dancing in the parking area of a luxury hotel; try performing a sonic flashmob around a gated housing area for oligarchs or around any government facility. Or try just jaywalking and dancing as a humanoid alien not apparently belonging to the hegemonic class of consumer citizens in an area provided just for this hegemonic class of consumer citizens: all of these unacceptable actions mentioned are considered close to sabotage, a crime, a blasphemic act if not terrorism these days. Being one of these provocateurs one would surely end up very quickly in the well-known wards, jailhouses, and prison camps, the reservations or hermitages for the outkast, the rejected, the criminal, the crazy, and the weird. Listening and sounding under the conditions of a governing dispositive of ubiquitous control becomes a precisely assigned and defined function.

Controlling sound

How does controlling sound actually serve the pervasive surveillance processes of admission, presence, and documentation? Dominant dispositives of a historical period or of a cultural context are not without ramifications and radiations. Any dispositive of social exchange, of trade and commerce, of cultural production, or of governance and political discourse is not disconnected from its host society. Hence, it is also a matter of interpretation why *this* specific material dispositive gained such an impact and importance for *this* culture under *these* specific circumstances. In the case of the pervasive dispositives of control on all levels of everyday activities of consumer citizens it is especially disturbing—from the concept of so-called *market economies* and *liberal democracies*—and, at first sight, rather surprising why a dispositive such as the one of ubiquitous control could develop, gain traction, and even flourish. Was there an underlying obsession for control all along that simply had not met the relevant means of expression in earlier states of these societies? Or was this urge for constant and consistently documented and exploited control always present in these societies? And, moreover, in respect to the issue of our inquiry: Why was it possible at all that exactly the obsession for sounds confirming the control of *admission, presence*, and *documentation* became one of the most important tasks for sound designers at large? Why is it that one of the major tasks for sound designers is—aside from supporting the dubbing and postproduction of movies and TV-productions and the invention of specific sound designs for film sound and game sound artifacts—the continuous invention, promotion, and further development of sound design in functional environments of traffic, banking activities, financial transactions, of IT-service, of workplace machine control, and, obviously, of the manifold mobile, wearable, drivable, inhabitable gadgets of distributed and ubiquitous computing that envelope your and my life at all times? Is designing sounds now almost in danger of becoming just another servile subdiscipline in the continuously expanding area of surveillance interface?

This form of control and its confirmation or rejection sounds for actions to be carried out by consumer citizens regarding admission, presence, and documentation is rooted in a long history of control architectures. Comparably pervasive control chains can be found obviously in the recently dominant concepts of project management flowcharts, in the mass production line, in earlier control and command chains of military entities, and, going deeper into history, also religious and tribal hierarchies demand similar, continuously confirmed and documented lines of control. Such lines of control are obviously no new phenomena in cultural history. Yet, it marks a surprisingly untimely appearance, as one could have guessed that similar dispositives might also have vanished with the advent of highly individualized societies with the urge to present, to praise, and to promote the foremost individual freedoms, freedom of expression, pursuit of happiness, and the multitude of individual sensory personae. Moreover, after the intense demands for artistic and so-called creative freedom since the secessions from art academies—also partially from conservatoires, and surely from religious art and the art of the royal court—a new established salon exhibition dispositive did emerge in the nineteenth century and was solidified in the various globally scattered movements of the avant-garde and post-avant-garde of the twentieth century: one could have imagined that the expansion of artistic freedom, of design extravaganza, of more and more individual creations in all areas of life and in all cultures of the globe could have been a byproduct of the ongoing global westernization and transcultural exchange of trade, production, and everyday life. But all of this might just be a self-indulgent and privileged perspective that ignores the incessant presence of slave markets, of oligarchic cultures, and of the predominance of market driven art's funding. Though the use of buzzwords such as *creative* and *individual* is indeed thriving globally, this might just heighten the sense of a certain paradox if not a weird contradiction between public lip-service and actual governance, between claimed policies and actual workplace conditions as well as the conditions of the public sphere. So, why did this *obsession with control* thrive and even take over the field of sound and music production?

Apparently, there is not one simple answer to this question already troubling our research from the beginning—starting at least with Chapter 3 in this volume: "Suffering: On contemporary sensory deprivation." Did we find in our research any hints and threads to at least better explicate and more complexly describe how this kind of suffering under circumstances of pervasive control could have become a dominant dispositive at all in the twenty-first century—and especially enacted in the area of sound design? Looking back on the field research, the interviews and questionnaires, the participatory auditions and observations of our research, it very quickly became rather obvious how *sonic labor* is forced to focus strictly and mainly on given architectures of control implemented in the contemporary dispositive of sensory labor in general. This begins in the control structure materialized in the "Location and Apparatuses" (Chapter 5) dominating sonic labor as for instance the hardware- and software-suites and the locations that also need certain login, logout, process documentation, and clearing structures. This continues in the expected and almost demanded "Skills and Habits" (Chapter 6) that sound designers of all kinds need to perform in their work environments in order to receive the relevant sound briefings from their clients and to deliver their products in the precisely requested format at the precisely defined time: admission, presence, documentation. Hence, the "Conflicts and Heuristics" (Chapter 7) dominant in the workday of a sound designer reflect precisely the urge to negotiate the demands from the client to the actual requirements of designing and generating certain appropriate and efficacious sounds. Even the "Portfolio and Presentations" (Chapter 8) represent and mirror exactly the urge of control and of a limited time, format, and space for *admission, presence, and documentation,* also in these case studies these are the three major categories for control. As a consequence, it seems to be necessary to excavate the details of sonic labor as a form of farming and harvesting sound as will be done in the next chapters dealing with the functionalization of listening and sound and the specific characteristics of sonic labor.

The observation of this pervasive sort of control in all areas of sound design and sound production only stresses the urgency of this somewhat unsettling fact: Why could it be that this discipline of sound invention, creation, and performance could degrade to some sort of support act in order to establish a frictionless functioning of an expanded surveillance apparatus? Or could it even be that this business itself does not yet realize where its function in the contemporary politics of surveillance lies? It seems that many protagonists and designers in this field still cling to the idea of their field as an arena for performing a magic trick—as explicated in Chapter 2 of this book: "Consuming: The brief history of a magic trick"—and they have not yet taken note of the aspects of "Suffering: On contemporary sensory deprivation"—as explicated in Chapter 3 of this book and "Ideologies of efficiency" (as also discussed in Chapter 3) that actually dominate this field—and considered that they might lead to "Imploding idealizations" (the last section of Chapter 3). If this obvious and hegemonic obsession with control is consistently ignored, overseen, negated, and never critically reflected, this might result in the fact—no longer so unreal—that sound designers who work for such a sonic surveillance state become the collaborators of an explicit or more implicit authoritarian government. How could a sound designer individually deal with this threat or danger? And what means would a sound designer have to be disobedient? Are you capable of just rejecting any job offers related to companies providing surveillance infrastructure? But, how can you be sure that the company you work for is *not* involved in such activities, if the surveillance dispositive is actually hegemonic? The questions arising from sound design's situatedness in dispositives of control are manifold and they threaten all concepts of creativitiy, personal freedom, even of business development, and, obviously, of all aspects of personal freedom.

It seems as if this development is rooted in much larger and much more disseminated and scattered developments definitely not restricted to the field of sound culture, of design labor, or of functional sound design in its narrowest sense. The contemporary culture of sound design apparently here is not much more than a symptomatic area, in which disturbing developments pervasive in contemporary societies become audible and sensible. Sonic labor is deeply affected and structured by these developments, in its minuscule everyday activities of drafting, presenting, refining, explicating, and implementing a sound design. Following the political economy of music, of noise, and of sound as proposed by Jacques Attali over forty years ago, one must concede today that the wild and rebellious energy in all the vibrations, reflections, resonances, and oscillations of sound and music apparently lead to a divergent impulse: An urge for containment, for control, for implementing and functionalizing musical and sonic and noisy practices into the governance architecture of society; be it by means of a militarily organized conservatoire and orchestra structure (as analyzed by Attali), or be it by means of connecting sound production with the production and supply chain of the entertainment or the media and surveillance industry. Sound apparently urgently needs to be controlled, it needs to be subjected under the strongest forces to support existing desires of control.

The silencing dispositive

How does a silencing dispositive in general serve the pervasive surveillance processes of admission, presence, and documentation? In order to control sound, it seems necessary to control all actors, their practices, all machinery, their functioning, and all possible emanations of sound. This control is executed by means of an overarching and socially as well as technically ordering dispositive. The dispositive that is dominant in recent sound cultures since the Industrial Revolution of the nineteenth century, its dominant acoustic theories and its pervasive expansion of sonic labor on all levels of society and economy, is apparently: *The Silencing Dispositive.* What are the constituents of this contemporary silencing dispositive and how do they affect the invention, creation, and implementation of new sound designs in particular and contemporary sound culture in general?

The silencing dispositive is a direct consequence of a culturally legitimized obsession with controlling sound and its function in the pervasive chain of controlling admission, presence, and delivery of documentation by the consumer citizen. This threefold control chain—very present in any diagram of signal transduction or any flowchart of process optimization—can be executed and controlled in the best way if the signals, documents, and the artifacts being controlled can be recognized, can be scanned, and can be tracked without any major distortion or ambiguity. The sound design projects we encountered, documented this strong will for complete explication and documentation: the urge for explicitness in their working models, sonic theories, design documentations, presentation rhetorics, and portfolios show exactly this will (*cf.* Chapter 6, section "The client whisperer"; Chapter 7, section "Pitching detour"; and Chapter 8, sections "Staging airwaves" and "Selling the sweat"). All of this actually amounts to a pervasive exclusion of ambiguity that represents one major characteristics of contemporary business and negotiating rhetorics—and as such also affects all neighboring crafts, such as sound design. However, such an exclusion of ambiguity, of distortion, and also of diversity, results directly in a process of *radical silencing* of anything outside such strictly predetermined categories: *categories of control limit the population of the controlled entities.* Not anyone can get in, not anything can claim presence, not any artifact is regarded as a valuable document for the activity of a consumer citizen. Hence, this dispositive silences the activities of unwanted or rejected consumer citizens, their admission, their presence, and their documentation. The process of silencing is therefore equivalent to a process of—speaking in terms of signal processing—receiving and filtering out a clear and detectable signal, apt for analysis and further processing. The signal *shall*, the signal *must* be crystal clear in this case. Anything short of extreme clarity is sorted out as useless and irrelevant if not criminal and illegal. It is a *radical gating* of access, a gating of citizenship in the end. Surveillance agents are audiophile signal transduction aficionados (see also Szendy 2016).

This urge for silencing lets controlling sound and controlling consumer citizens seem possible and desirable. But where can such a radically harassing and limiting urge draw its cultural and historical legitimation from? Is there any legitimation of such a practice of pervasive silencing—or is it just a historically and culturally established practice, an ugly habit that is passed on from generation to generation? One perspective, from sound studies and history, can be found in Jennifer Lynn Stoever's investigation on how sound and listening did actively produce the racial politics of today. Her book, *The Sonic Color Line,* "presents a cultural and political history detailing when, why, and how listening became a racialized body discipline and how it both informed and was informed by emergent sound technologies" (Stoever 2016: 4). In such studies the minuscule sounding and silencing actions as well as the sonic nanopolitics (Schulze 2018: 136–166) countering different approaches are excavated.

In cultural history, one surely can never completely rule out the impact of unreflected common truisms; but one of the distinct roots for such silencing dispositives can be found in the fabrics of modern societies—including discursive regulations in cultural production, in the sciences and the arts, in education, and also in governing in general. The modern bourgeois cultures, established in the eighteenth and nineteenth centuries and drawing from references, authorities, and reimagined antique traditions did single out certain types of consumer citizens as their founding personae, between lawmakers, entrepreneurs, artists or philosophers, scientists or warriors, inventors or orators: they did not rely only on the mass precariat in its disturbing heterogeneity as their founding nonactor. The silencing dispositive is therefore not external or alien to contemporary societies: it is intrinsic to them. These collectives consistently apply forms of silencing in their everyday activities, and they are fundamentally founded on its dispositive in order to maintain established hierarchies of power and exclusion of groups possibly distorting their execution of power. *Clearing the signal* is—in regard to power structures—basically a means of maintaining or gaining power. Allowing distortions to occur therefore equals a basically democratic framework intending to allow continuously more sources and voices to articulate themselves—even if untrained, unusual vocal habits, sonic personae, and strangely alien sound events take place. This erratic quality might be generative. In a silencing dispositive, though, another strategy for generativity is implied.

Generativity after controlling sound is thought to operate by an evermore subtle chain of refining, filtering, and processing operations. The quality of sound is—that is the main idea here—already stored in the original distorted and contaminated signal or entry, it just needs to be purified, to be cleared, until the original quality and generativity can be accessed. The essentialism and the imaginary purity implied in such thinking is hard to overstate. This figure of thought, dominant in all silencing and filtering dispositives—and as such in all discourses on precision in design or in the arts—adheres in a quite unreflected way to an ancient biblical if not Catholic concept of *original purity, holiness, and an immaculate state* in all things and beings: an originally immaculate state that has only been impurified, destroyed, and sinfully contaminated by subsequent actions and occurrences following *The Original Sin.* Any concept of signal processing and signal filtering, actually still follows this Christian idea of an essential purity that needs to be filtered out—and that *actually* can be filtered out. *Signal processing is a Christianist belief.* The dispositive of signal transduction and noise reduction, consequentially, expands this idea of an original purity into a whole scientific paradigm with its subsequent media technology and media entertainment industry. This silencing dispositive is built into all contemporary technologies and cultural practices. From this foundation, all enunciations and instances of *noise*—be it in persons interjecting, groups of people complaining, or agglomerations of humanoid aliens and their apparatuses performing weird sonic patterns—must be regarded as major insults, major repercussions of *The Original Sin,* and therefore a major blasphemic attack on the belief system of silencing, purifying, and filtering. He who complains about noise, implicitly assumes for Himself the position of *The Infallible Lord.*

From this major auditory belief system that can be recognized in contemporary surveillance societies there remains only one position for actors and listeners and sound producers of all kinds: it is the position of *Audiopietism* (Schulze 2007; Papenburg and Schulze 2016: 3)—encountered especially in Part Two "Sonic Labor," regarding the urge of sound designers for minimalism, for sonic excellence, or for a perfect production. A common expression of this belief of *sonic transcendentalism* and metaphysics (briefly touched upon in Chapter 4, section "Drifting, sensorially") can be found in the pitch presentations or public talks of various sound designers, sonic thinkers, sound artists, composers, or audiophiles: "Music is the most powerful form of sound that we know that affects our emotional state" (Treasure 2009: 1:20). This statement might qualify as thoroughly empty and void of meaning from a research perspective as it can neither be verified nor falsified by any research method one could imagine. However, a statement like this is very common to signal one's social affiliation to a certain group of sonic *pro*sumers who believe in the purifying, in the optimizing and the self-improving quality of listening and sound design. The core of such truly precritical, audiopietist claims, which can be found in any educated consumer citizen's common knowledge about cultural practices in music, is another tedious iteration of the age-old "audiovisual litany" (Sterne 2012): the belief of *audiopietism* is centered around the assumption that if only one would devote oneself to listen more carefully, to enjoy more silent and more meticulously crafted sonic artifacts, only in sound environments of the highest quality assigned by contemporary audio technology research, and if one would then reorganize one's life around such *audile techniques* (Sterne 2003a), strictly assuming a pure *hearing perspective* (Auinger and Odland 2007), then one would become *A Better Human* without fail. The fallacy in here is obvious yet also well concealed. The individual constituents of this belief come from audio technology research, from audiology, phonology, and from psychology, even from science and technology studies as well as from sound art; they represent all in all well established and reflected positions in themselves. However, it is a simply incorrect and unacceptable misuse to draw such major and actually insulting, moralist conclusions from these careful and cautious scientific and artistic approaches. All of these approaches simply do not imply such a moralist verdict of this kind. Effectively, the most sensible and critical accounts of musical or sonic anthropology are boldly and distortingly extended to these horrifying teachings of audiopietism, a kind of *sonic eschatology*. Universal salvation and redemption is expected to ooze out and magically transform one's existence from an imagined area of sonic wisdom and perfection. The silencing dispositive requires such a metaphysics and it stabilizes a highly ideological and anti-materialist concept of sounding, of listening, and of all performativity around the sonic. As its most unsettling consequence, this ideology of audiopietism allows for any exclusion and any verdict against any group of humanoid aliens not apparently belonging to the hegemonic class of consumer citizens. Public discussions around noise regulations (Bijsterveld 2008; Thompson 2002), preferably framed as merely technical discussions, soon deteriorate to expose their actual goals: moralist verdicts are released, lifestyles are condemned as inappropriate, cultures are excluded as too alien, too unbearable, simply superfluous and annoying. Noise regulation very quickly turns out to be cultural regulation, exclusion of the rejected and detested. Rebellious youth cultures, immigrants travelling over borders, or unwanted lifestyles are gated. The dispositive therefore fulfills its mission as soon as it is used against any form of unwanted sound: the mere notion of unwanted sound is hence in almost every case to be explicated by the notion of *unwanted cultural practices, lifestyles, forms and expressions of existence*. The exclusion of the rejected noise of other humanoid aliens is part of the effort to eradicate those traces and actors not apparently belonging to the hegemonic class of consumer citizens. It is a *sonic cleansing*.

Functional listening

How does functional listening serve the pervasive surveillance processes of admission, presence, and documentation? The silencing dispositive, its cleansing activities, and its subsequent, audiopietist forms of measuring, evaluating, transforming, producing, and postproducing sound require a specific listening mode related to its initial assumptions. These assumptions culminating in the highly refined concept of the crystal clear and cleansed signal demand from any listener something that is definitely *not* common in everyday listening. This is *The High Mass of Audiopietism*. Hence, in detecting, measuring, and analyzing recorded traces of sound for audio technology purposes the distinct and clear signals are of major importance—but they are not necessarily so in the course of a day, in the times of one's life. Blurriness, indirect washiness, layered and unstable signals are on the contrary often a source of major joy and excitement, inducing a sense of a specific situation and place, markers of a very recognizable, of a memorable and dear moment in time, of a location in space. I love it, when the blurred sounds of songs I love can be heard filtered and blurred outside of my balcony. A crystal-clear signal would in these respects resemble an erasure of all of these markers, these moments or situations. It is an exclusion of not intended activities, an *authoritarian cleansing of the sonic environment*. Yet, the silencing dispositive, once established, technologically implemented, and ideologically supported by audiopietism on various levels of research, transmission, society, and politics, does not grant major deviations from its main assumptions. Such deviations pose a serious threat of destroying its whole purpose of providing a common ground for the exchange, the analysis, and the further use of signals; at least for all humanoid aliens belonging to the hegemonic class of consumer citizens.

This *radically reduced form of listening* and of sound design is a prerequisite for contemporarily established forms of sonic labor and sonic economies: in this very reductionism and stereotyping of sonic patterns, the long histories of colonial and territorial overreach and abuse are being inscribed in sound design as markers of exploitation and oppression (*cf.* Chapter 14 and Chapter 12, section "Sonic stereotyping"). You are not allowed to articulate what you *sense* while listening—you are expected to connect bits of your sensations to hegemonial listening discourses of cleansing, a crystal clear signal and audiopietist belief. Listening here *is* a mere function and not much more. And this function is even more restricted than one would imagine anyhow in the professional framework and the contemporary working environments (as presented in Chapter 5, section "Staging the startup"). This regulated form of listening follows the imperatives of controlling sound and is a direct consequence of the silencing dispositive. It intends mainly to restrict and to regulate all more disturbing, distracting, or dissolving forms of resonance and to secure one form of listening and of understanding. This form of listening is *functional listening*. It is a socially trained and a professionally required professional practice that is not to be confused with your or my everday listening practices in the kitchen, whilst commuting, whilst consuming media products, or when visiting locations for entertainment or recreation; these practices at least potentially steer us toward a joyful, desirable, and idiosyncratic form of listening and sonic experience. Functional listening can be found as a listening mode in the intersection of Barry Truax' *listening-in-search* (Truax 1984: 19) and Michel Chion's *causal listening mode* (Chion 1994: 24–28)—in combination with a kind of critical and distant listening (Nardi 2016), the clear and vocal opposite of the famous practices of *deep listening* (Oliveros 2005) or even *thick listening* (a term that alludes to the ethnographic method of a *thick description* with multiple, condensed layers of interpretation and narration; it is used foremost in the context of spiritual and religious listening and singing practices in order to address the multiple layers of experience, interpretation, and spiritual connectedness: Harris 1996: 8). These complex, fourfold similarities and oppositions provide a highly applicable frame for defining this functional specimen of listening. As in both first listening modes mentioned—*causal* and *in search*—the mode of functional listening is actively searching for and angrily ready to reacting to any sounds that do not hide the cause of their generation and their intention. They are explicit and defined sounds, distinct sounds, that can trigger and satisfy such a listening mode: sounds that are marked as irregular, as transcending given limits of audio production or of effects not expected to be generated in a given econocultural or sociopolitical context. A yawning, a hunch of hum, noises of other people or animals present, slowed down sounds of doubt and hesitation—or of inacceptable desire, excitement, stuttering, or a too rapid articulation of phrases and sentences. One therefore *critically* listens for flaws and glitches, hisses and involuntarily slips of the tongue; they need to be eradicated and deleted from the stored audio file—or they need to be muted and excluded from the audience hearing this transmission. Functionalized listening is hence the main practical expression and the main product of the silencing dispositive and its audiopietist desires of control. The sonic artifacts in question need to be subjected to the rules and standards of the silencing dispositive and it is your duty as a functional listener to detect these flaws and disturbances and noises that could possible hinder the smooth operation of the silencing dispositive. As a functional listener one decides from a critical distance to confine one's aural attention to be a functional servomechanism of this dispositive.

With this narrowing down, this restraining, this task orientation of listening, the most joyful, erratic, and wandering, easily distracted and also easily excited or bored moments of listening are quietly excluded. They are recognized as noise and disturbance—which they actually are in a societal framework of consumer citizenship. Functional listening is deeply annoyed and disturbed by joy or boredom, by individual affects regarding their listening objects, the sources they need to scrutinize and observe. Functional listening on the other hand enjoys the discoveries of perfection and of precise adherence to a given law, a given guideline, a given structural predetermination, a dispositive: you are a good entity in this system as long as you perform the expected behavior. It represents the lust for pattern recognition and for identifying learned sets of characteristics and patterns, again, from a critical distance. The sonic pattern in the strictest definition, hence, is the main operation field for functional listening. As soon as a sonic pattern is precisely defined by time-structure, timbre, and dynamics, it represents the perfect starting point for an exercise in functional listening. If a sonic pattern is not yet defined or only progressing toward a clear definition, it drops out of this process of functional recognition. It is rendered useless.

Returning to the latter two opposed listening forms mentioned earlier, *deep listening* and *thick listening*, these forms of listening are also not interested in sonic patterns in their materiality as such—yet in both cases they focus on often neglected, idiosyncratic qualities. The sonic patterns that are so crucial for functional listening as a material reference and a precise catalog of criteria to scrutinize, actually only partially contribute to the sonic experience that is at stake when learning, training, and performing these two forms of listening. In an ideal usage of sonic patterns they would all open up the generativity of listening and crafting—as the design pattern language originally was intended to be by Christopher Alexander (*cf.* Chapter 1, section "The dialectics of functionality"); and in ethnographic research of design processes, this concept still proves to be valid and generative (*cf.* Part Three, *passim*; but especially Chapter 9, sections "A sonic pattern language?" and "Performing sonic patterns"). Yet, in the everyday design practice—that represents one empirical core of our research—this generative openness is easily lost in the silencing dispositives and its ideologies of effectivity. What could then strengthen and enliven this generative character in design?

Deep and thick listening come together in the interest for the actual personal, also highly idiosyncratic effects that a sonic artifact or a sound performance can have on your or my sensory corpus—including all our personal, social, and intimate biography embodied in one's listening habit. However, both practices are distinct in how to follow up on these affects, reflections, and reactions. Whereas deep listening is actually a practice situated between sound art performance and sound meditation that finds its goal in a highly focused performativity, the practice of thick listening is less precisely defined. The term thick listening is often used by sound artists but also by scholars from education or religious studies. It refers mostly to an individual listening practice that intends to grasp most of the historical, social, cultural, also technological and biological ramifications of a specific sonic experience. It is different from an audiopietist urge for sonic transcendence as this thick listening (as also is deep listening) does not claim to achieve an immediate or at least causal optimization or self-improvement of the listener. It is not a therapeutical, purifying, or cleansing activity. Both forms of listening now lack the functional hyperfocusing and self-optimization that is characteristic for the scrutinizing listening practices attached to the silencing dispositive. To the contrary, if one performs deep or thick listening the whole range of sonic affordances and of erratic sonic experiences is acknowledged as the actual grounds and reference for listening. For example, coming back to an earlier example of transcultural sound design (*cf.* Chapter 12, section "Sonic stereotyping") one would then listen materially thick and deep into the actual—not just imagined and projected—concepts of culturally embedded sounds. As a consequence, the resulting sound design would at least partially explore and propose sonic patterns that were not stereotypically reduced to the lowest common denominator—but they would seducively also guide into the experiential sonic experience of a region and an everyday culture. Deep listening and thick listening as starting material exercises for sound design can indeed provide radically new and even surprising sound material—which has the power to attract audiences in a different way than just through a recognition of well-trained and clichéd sonic patterns. It would craft and invent and implement effectively new sonic patterns. Thick and deep listening have the potential of sensorily reflected, performatively critical forms of listening. But a functional, search- and causality-oriented framework seems more often than not to prevent the application of such techniques.

Functional listening, all in all is a form of listening that rejects in general such vivid and dynamic transformations. It prefers static, regulated, and patterned articulations of sound that it can scrutinize and subject to the cleansing, filtering, and purifying processes of the silencing dispositive. One can assume that any disciple of functional listening finds their great joy in detecting unfiltered and impure specimens in their sonic artifacts, and subject them meticulously to the then necessary process of cleaning up and achieving a higher ground of transcendent if not supernaturally immaculate states. The *obsessive-compulsive character* of such a form of listening is hard to ignore; and also the deeply *thanaticist desire* performed when listening functionally. This thanaticism, as described and explicated by McKenzie Wark, constitutes a major cultural trajectory for societies relying on activities of consumer citizens, processing along the lines of admission, presence, and documentation, contributing with these activities foremost to the production of exchange value:

> Thanaticism: a social order which subordinates the production of use values to the production of exchange value, to the point that the production of exchange value threatens to extinguish the conditions of existence of use value. (Wark 2014)

The approach of radical functional listening is an expression, a specimen, and a form to listen according to thanaticism. This maximum reduction to noncorporeal, product-oriented, and functional listening is a form of listening that at its core never focuses on individual actors and sound events but on its product. Functionality is—if driven to this extreme—bluntly thanaticist as such.

14

The Economy of Sound

A theory of sonic labor

The sonic workforce

Sound design in the early twenty-first century is a field of labor that structures and disciplines its sonic workforce through milking its ambition and driving it into exhaustion, so that a more sustainable form of regenerative labor in the economy of sound remains currently more a noble wish than a lived reality: this is the second major result from our fieldwork of sound design projects of differing scales and of varying applications as well as our critical analysis of the history, the design approaches, and the listening practices of sound design. Labor in the area of design is still in a state of strange ambivalence, a sort of obliqueness if not murkiness from an outside perspective—and almost over-theorized, terminologically replete, and densely articulated from the inside perspective. As in other disciplines, a line of distinction can be drawn in sound design between a publicly visible and highly recognized star personnel, and a rather invisible, rarely mentioned if not anonymized minion personnel. This division can be observed also in areas such as cooking and tailoring, of hairdressing and shoemaking, of interior design and furniture—and generally in almost all areas of design and the so-called creative industries. This distinction is not so much one in regard of actual design impact or of everyday experience, but one of historically rooted distinctions that are at the same time gendered, market-driven, and separating professional labor practices that seem to be identical at first sight. Hence, the major difference in these areas of activity and production lies not so much in the recognition of its protagonists—though one might be inclined at first to guess so—but in the actual conditions of labor (which extend obviously into public recognition, visibility, and accolades). Therefore, this present study does, most importantly, *not* deal mainly with star sound designers or composer sound designers. They enjoy a by far different and almost incomparable condition of labor that comes closer to the labor conditions of any living celebrity artist, composer, engineer, software inventor, or couturier. Their work has to be analyzed and understood as part of a globalized and highly dynamized market of luxury goods and commodities of maximum distinction.

In contrast, the theory of sonic labor that we develop here, however, is concerned with the quantitatively impactful form of sonic labor done by freelance sound designers, by sound design departments in companies and communication agencies, as well as the contracted teams and groups of sound designers working together just for a limited amount of time to work on a given sound design project. These are mostly not the aforementioned star designers or author composers, yet they are members of the supposedly infinite workforce, often anonymous and nameless, only sometimes credits are given to them. This *class system* inside the broad variety of all design professions is often ignored when speaking of *design culture, design research*, major *design works* or *design history* implicitly referencing only works of star designers—and speaking condescendingly of the large design workforce. In this study, this common hierarchy of design is challenged: the emphasis of our study lies on the everyday consumers of and everyday listeners to sound design, and the many ephemeral artifacts crafted in the vast area of sonic labor. The more prominent examples of sound design, though—created by famous composers, musicians, artists, or inventors—are not excluded from this theory; but they represent a rather specific and often highly exotic working process under refined if not luxurious working conditions. These famous examples entered therefore only as selected references used by sound designers in our empirical study on *Sonic Labor* (in Part Two, Chapters 5–8)—but they did not form the main corpus of our study. There might be connections and overlaps between both conditions of labor—but not necessarily so.

The *Sonic Workforce* as it is constituted in the 2010s is largely immersed into the labor conditions that apply to the so-called *creative class* (Florida 2002; Montgomery 2005) or—with the recently commodified governance buzzword—*creative industries*. Whereas the concept of *creativity* needs to be deconstructed as an unsuspectedly successful byproduct of early twentieth-century workplace psychology in its largely instrumentalist and utilitarist appearance, this was not the major goal of our study. Instead of a deconstruction of terminology, epistemology, and historiography of the concept of creativity, this theory proposes a materialist critique focused on the actual, minuscule conditions of labor in the field of sound design (*cf.* Chapters 5 and 7). This critique might—by extrapolation— then, in not only a few respects, also apply to many other fields of contemporary and future sensory design. This theory of sonic labor might therefore serve as a blueprint for theories of sensory labor in general. To recapitulate, following our fieldwork into sound design projects and our critical analysis of sound design approaches, what is *sonic labor*?

Sonic labor can be defined as the professional and highly affective activity of a sound designer who is at the same time a sonic consumer—a consumer citizen. This protagonist is part of the sonic workforce as it is contributing in both of the aforementioned roles as sound designer and sonic consumer to a workflow organized in order to provide certain requested sonic artifacts to a larger media product in soft- and/or hardware. A theory of sonic labor therefore needs to connect this major area of economical and societal activity related to sound to other areas of everyday life, of sensory experience, of contemporary policies of control and austerity, and—last but not least—to concepts of the listener, the musician, the sound artist, the sound aficionado, or the audiophile. The sonic workforce includes in the 2010s largely—if not only—*sonic prosumers*: however, this once maybe liberating if not futurist concept, intended to transcend limited and outdated concepts of labor and of craftsmanship, proves to be more of an ambivalent if not depressing concept of ignoring if not willfully degrading craftsmanship and labor rights as part of expanding workforce control into every single nonwork-related aspect of everyday life since its introduction by Alvin Toffler in 1970 (*cf.* Chapter 5, sections "Private workbench" and "Suites of production"). The *sonic prosumer* (Toffler 1970) is the ubiquitous (Kassabian 2013) and 24/7 (Crary 2014) slave of *affective labor* (Hardt 1999).

This sonic workforce as the driver and the substance of sonic labor operates in a largely invisibilized area of production—though the actual effects and also the production of these effects are not at all invisible: their designs are heard a million times every day, they have a tremendous impact on everyone's daily rhythm of working and recreation, they result in various work-related symptoms and pathologies or coping mechanisms. The designers' lives as well are reversed and restructured according to current projects and the often violent demands and self-harming work habits of project partners and employers (*cf.* Chapter 7). The effects of sound design and of designing sound therefore are material and physical, they demand a massive physiological, sensological, and affective investment by their protagonists. The forms of exploitation though are highly refined, specified, often concealed, and to a large extent invisibilized themselves. By this invisible exploitation sonic labor itself contributes to the silencing dispositive of societies and economies of control, thus maintaining the continuous, embodied, and pervasive control of sounding and listening, of sensory events and their experience (*cf.* Chapter 13). Sonic Labor is on the other hand also one of the major drivers of expanding sensological approaches territorializing and colonizing your or my very own everyday life. What once was only a small, marginal discipline of sensory design for the hearing senses has become in the early twenty-first century almost a fundamental first step into designing, organizing, and filling all the other thinkable sensory sensibilities of humanoid aliens: *your sensory corpus is now up for grabs!* And in the course of the next decades, side by side with the sonic consumer there will appear *olfactory consumers, kinaesthetic consumers, tactile consumers*, or even *proprioceptive consumers*. One can be quite sure that *every* sensory area of humanoid aliens like you or me will be exploited, commodified, and monetized. The area of sound design and sonic labor is just the exemplary precedent, just the first step. The future founders, developers and major entrepreneurs and artists in these new areas can already learn from the mistakes and the clever market moves of sound designers from recent decades.

Ambition and exhaustion

How is the sonic workforce being exploited by milking its ambition and driving it into exhaustion? The affective and sonic labor demanded from the sound design workforce relies to a large extent on two *affective resources: ambition and exhaustion*. Both resources are—from the perspective and conceptual framework of material culture and sensory studies—definitely not a case of *immaterial labor* (countering Hardt 1999: 93ff.). They represent a strand of sensory and generative labor that uses its producing personnel, its workforce, as its physiological and sensorial resource. To make this perfectly clear: the sonic workforce is not only a workforce in the traditional sense, providing the material and physical and skillful activities in order to invent, execute, and finalize a product in the production process; but they are, moreover, also the resource from which concepts for designing new sonic patterns, finding relevant sonic affordances, and crafting new sonic artifacts are taken from. They are the miners and the mine at the same time. In auto-mining their own skills, sensibilities, their knowledge and their aesthetic sense, they perform a kind of *generative labor*: they generate, they produce, they bring to life almost all by themselves. In this sense, the sonic workforce resembles more strictly artistic or performative professions; with one striking difference: the means of negotiation, of planning, of distribution, and of implementation are mostly not accessible to themselves but only to larger companies and agencies. The previous examples regarding sound design workshops serve as proof for this. It is precisely this concept of *generative labor* that might lead us later to subsequent far-reaching conclusions.

This still new specimen of labor has been colonized since the "third spirit of capitalism" from the 1970s to the 1990s (according to Boltanski and Chiapello [1999] 2005) by an approach to labor that in earlier decades and centuries was mostly known in the domains of *desiring-production* and *caring labor* (Hardt 1999 referring to Marx and Freud). These days, however, most specimens of sonic, of sensory, and all sorts of design or *generative labor* obtain the characteristics of caring labor and of desiring production: the production and the invention of highly desired and craved, of addictive and drug-like artifacts is the domain of design labor; at the same time this production of addictive commodities makes it necessary for their sonic prosumers to be highly versatile in all sorts of caring labor in respect to their clients, their business partners, and also their team members or relay contacts in the production chain (*cf.* Chapter 6, section: "The client whisperer"). This continuous affective labor is materialist and exploitative to a degree that equals in part a process of progressive self-annihilation. The deep joy of effectively disappearing in the final sonic artifact is a self-observation (and a self-critique) many of the sound designers in our research shared with us (*cf.* Chapter 7, Chapter 6, section "The client whisperer," and Chapter 8, sections "Grey minimalism" and "Channeling the shaman"). Creating, for instance, the main sonic pattern for a sound branding or a set of signal tones, works as a mark of individual achievement, which is a satisfactory task in itself; though, it won't pay the rent. But how is this satisfying joy now connected to the working process in sonic labor? How do *ambition and exhaustion* as major resources for sonic labor drive this process of generating new sound designs?

Ambition as a driving force of affective labor is the dark, the vantablack core of this specimen of labor so to say. On the surface and in all the official statements and press releases, published interviews and designers' statements on their work it could often seem that a mere and intrinsic desire for aesthetics and design, for form, creation, color, and tone are the driving forces of sonic prosumers these days. Yet, obviously—and showing in their actual actions, their forms of presentation, and their efforts for institutionalizing this design discipline—another, even stronger and maybe way more important driving force is probably ambition. Ambition is the burning and plastic, the visceral and a vulnerable core of the desire in generative labor. If the desire is to create and to publish, then the desire to get recognition, to be praised, and receive accolades is the most material resonance one could get. In the first stages of such a career, money or any material revenues are often publicly disregarded as only necessary for living, but actually ignorable and irrelevant: honorariums and profits are in general ridiculously small if not nonexistent. Hence, ambition grows into the core driver of sonic labor. As a sonic prosumer, this is exactly what you seek, where you get your energy and your motivation and what keeps you doing, acquiring, and completing your individual jobs—even if you feel horrible, miserable, exploited, and mistreated if not abused while doing so. Your ambition is what keeps you going. And the actually life-supporting job on the side or the support by parents, social security, or inherited wealth that brings home the notorious bread and bacon is usually hidden from sound agencies or clients; wouldn't it seem awkward and miserable if they knew you would need to work as a waiter, in a call-center, or be dependent on your parents or your wider family just to keep your actually desired job? No client would ever think about raising the honorarium if he knew this background information. A client might just sack you for being so irrelevant and without success or recognition, right? Ambition as the driving force might be of use and a useful coping strategy at the beginning of entering the sonic labor workforce, but in later stages of a professional career it might turn into a rather dangerous vulnerability.

Exhaustion of any kind is, consequentially, excluded as a mere reality from this specimen of labor: it is the physical, individual, and also idiosyncratic limit to any ambition, to any generative labor in general—and to sonic labor in particular. When ambition and its side-effects are hidden, then exhaustion turns into an even more ambivalent, dangerous, and truly life-threatening experience. Professional exhaustion is, on the one hand, regarded as a symptom of a full portfolio and hence more of a badge of honor to prove one's efforts, one's successes as well as the incessant demand of clients, team members, colleagues, and future contractors; on the other hand though, there is a fine line between being requested and recognized and grinding down in an almost slave labor-like series of endless sessions in front of the sound producing set-up of software suites, sound generators, instruments and recording devices (*cf.* Chapter 5, section "Suites of production" and Chapter 6, section "The client whisperer"). This latter reason of exhaustion is one that stands in direct conflict, so it is perceived, to the genuine ambition and the joy one has in designing sound: *if you like this very profession that much—why could any task we give you ever be stressful, annoying, or exhausting for you?* It is exhaustion and despair, the feeling of being worn out, pressed out, and abused that makes a difference in classes, lifestyles, and life expectancy. The designer who can rely on outside resources for a living (lucrative money jobs or inherited wealth) will not suffer too much of this kind of exhaustion. One might recognize here the distinction between a *hermeneutic class* from an *exhaustion class* (Lippe [1987] 2000: 295): whereas the *hermeneutic class* fulfills their work duties mainly by comparably effortless sign operations and sweatless writing processes with a massive impact, the *exhaustion class* has no other option than to exhaust themselves and to wear themselves down by selling their physical power and concentration. The latter class gets to be bodily sucked out until they die, the former class accumulates even more energy, capital, wealth, and all forms of access. These are the frontlines of *semiotic capitalism* (Szepanski 2014b).

All these intrinsic dangers and inner contradictions in areas of affective and generative labor are hard to fend off. They pose a continuous threat to all designers and artists working in these professions (Hesmondhalgh and Baker 2011: 226–228, *passim*). Yet, this is not only about a difference between so-called "good and bad work" (25–51, 220–234), but about the actual, material culture of sonic labor and all areas of generative labor. The main issue here is not just "how to incorporate creative labour into consideration of the distribution of good and bad work across societies as a whole" (234), but how these forms of labor are actually, materially and in the very meticulous details of location, of apparatus, of skills and habits, of conflicts and problem solving strategies, of presentations and portfolios, realized and actualized in your or my life. The highly ambivalent and questionable glory of the so-called *access economy* (also known under the more propagandistic names of *on-demand, collaborative, sharing*, or even *gig economy*) is just a foreshadowing of future colonizations and monetizations of the most remote areas of every single individual life: *your very moment now, no matter where you are or what you do, why not monetize it and feed yourself, even here and now into the processes of producing exchange value?*—Gotcha at your ambition, right? And your exhaustion will surely follow.

A labor theory of sound design

What are the constituents of the field of labor in sound design? Sound design as sonic labor is constituted out of four major material and performative layers: *(a) location and apparatus, (b) skills and habits, (c) conflicts and heuristics,* and *(d) portfolio and presentations*. These layers are the main playfields where individual ambition and exhaustion are actualized and experienced. A theory of sonic labor explicates on these playfields the minuscule, everyday activities of sound design that craft new sonic artifacts consisting of new sonic pattern languages as well as new sonic affordances. As the actual habitat of the sonic workforce these four fields allow us to excavate the material culture and the detailed performativity of sound design practices as visualized in Part Two of this volume: "Sonic Labor" (Chapters 5 to 8). On the following pages, the details of this habitat are unfolded and explicated with regard to the major insights resulting from our fieldwork and the interviews with sound designers.

(a) *Location and apparatus* (Chapter 5) contribute the material shell and the physically visible conditions for sonic labor. They are apparently bound regarding location to the highly local and arbitrary structure of housing, gentrification, accepted working conditions, and contemporary working culture (*cf.* Chapter 5, section "Staging the startup"); regarding the apparatuses used, displayed, and acquired for sonic labor, they are constantly provided by the collective process of negotiating the standard of sound productions and the individual, at times idiosyncratic, preferences, personal memories, and corporeal inclinations toward a certain instrument or generation of sound production soft- or hardware (*cf.* Chapter 5, sections "Private workbench" and "Suites of production"). Actual sound design practices are, obviously, not thinkable without these material requirements. *The locations in which sound design takes place and the manifold apparatuses that are factually used to produce sound designs are always limited; they guide the individual protagonists of the sonic workforce into an even more restricted array of production styles.* The characteristic locations presented in Chapter 5, "Location and Apparatus," of this book are architectural dispositives; they represent preferences that are not innocent. In their focus on highly personal locations (personal studio, apartment, personal space in a public café or in the backstage of theater or concert venues, Figures 5.1–5.5 and 5.18) and personal set-ups (focusing on personally accumulated musical instruments, cloud-based software and sound storage, or large and preferably analog equipment, Figures 5.6–5.15) they foster effectively the aforementioned character of *affective labor* to develop in the work of sound designers in the 2010s. Their labor of generating new sonic artifacts is, concluding from these material enclosures, considered by their clients or by their contractors as a personal activity. Sonic pattern languages are invented and new sonic affordances are discovered in these personal habitats: personal findings. Nevertheless, a professional behavior and a timely delivery of documentation according to the control structure of *admission, presence, and documentation*, discussed earlier (Chapter 13) is strictly expected. All these personalized details of this sonically affective labor though allow for an even more demanding work schedule with at times moralistic if not threatening undertones.

(b) *Skills and habits* (Chapter 6) manifest the embodied and sensorially as well as habitually intrusive effects of continuous and intensive sonic labor on the sonic persona of a sound designer and their practices (*cf.* Chapter 6, sections "Sonic imagery" and "New ear"). The demand for these skills results mainly from actual sound design projects as well as highly personal inclinations and desires to know in more detail the functioning of one sound effect, production technique, or magic sound trick (Figures 6.1–6.5). These skills are in part learned in so-called officially certified courses, workshops, and seminars and in part learned through the usual, continuous and casuistic self-improving that most freelance designers need to follow; online tutorials, new literature, basic manuals as well as intensive personal training by a colleague who is punctually more experienced in this or that production technique (Figures 6.11–6.15). This mixed and highly situated learning and teaching process blurs consequentially the lines between the professional sonic workforce and the rather uneducated or less educated amateurs in it: they are equal in their function as sonic prosumers. *The skills that these sonic prosumers acquire as well as the habits they learn to inhabit over time are largely coping strategies regarding clients' and contractors' demands for highly specific outcomes of a sound production process.* These coping strategies (Figures 6.16–6.20) again stress the aspect of affective labor and *caring labor* regarding the needs and desires, wishes and obsessions of their clients. As a sound designer you are the nanny if not the personal counselor of your client. You may not wish to be, but you actually are: the specific materiality of sound, its affective load, and its materialist effects guide clients to heavy expectations and emotional reactions, often derived from notions close to the sonic metaphysics of *audiopietism* as a *theology of sound.* Personal and intimate habits (e.g., personal lingo, Figures 6.6–6.10) regarding sonic affordances and sonic materialities and how to work with them result in part from these outer demands and expectations, intending to counter or, punctually and strategically, to foster them.

(c) *Conflicts and heuristics* (Chapter 7) are the cultural forms in which highly specific and nonroutine encounters between individual designing desires and the professional work conditions are manifested: they result in the aforementioned refined coping strategies as skills (*cf.* Chapter 7, sections "Ephemeral lie" and "Sonic darlings"). Conflicts are the actual material of a sound designer's affective labor on an everyday basis and the individual heuristics develop with these work practices: "How do I deal with the fact, that this client or that team member does not agree to this project outline or is not capable of contributing to this? And how do I still steer our project rather smoothly through this sound design process?" (Figures 7.6–7.10). *The conflicts between clients and sonic prosumers as well as the heuristics those prosumers learn to use represent encounters with the silencing dispositive incorporated in the economy of sound and the prosumers' individual efforts to craft workarounds not to be grinded down to obedient sound controllers.* As the silencing dispositive demands a sonic cleansing as well as an uninterrupted, weightless, and noiseless production process (in conflicts with Figures 7.1–7.5, 7.11–7.15) this stands in clear and permanent contrast to everyday tasks of bringing team members together and satisfying the, at times, childish needs of a client (*cf.* Chapter 7, section "Pitching detour"). Most of all the urge for sonic transcendence and audiopietism is a source for conflicts between clients and sound designers. The individual musical biographies, intimate, sonic resurrection experiences, and the resulting sonic personae of clients might lead them to essentialize and generalize their idiosyncratic needs and pleasures; but the professional practices and the large experience of sound designers with various sonic affordances and their sonic pattern languages provides a rather reliable ground for work practices in reference to the actual forms of listening of the intended sonic consumers (*cf.* Chapter 7, section "Design abuse"), transcending yet incorporating mere sonic darlings (Figures 7.16–7.20). An often hardly resolvable conflict: the personal, artistic designer has a knowledge base of sonic consumers that a client wants to address; yet the client themselves indulges to a large extent into not really reliable but individually strong affects—ignoring willfully sonic consumption, prefering to cite truisms of sonic metaphysics that have no actual meaning for sound production and its material manifestations (Figures 7.1–7.5). In the end, the client wins, but what he wins is often determined by the sonic prosumer (Figure 7.9).

(d) *Portfolio and presentations* (Chapter 8) are highly formalized and regulated public articulations regarding individual work processes of the sonic workforce and their individual results (*cf.* Chapter 8, sections "Staging airwaves" and "Selling the sweat"). They result in part from the expectations of potential clients and sonic consumers as well as from the material settings provided by media distributors as well as hard- and software developers. Yet, they also represent in trace elements the demands of a silencing dispositive and even the ideology of audiopietism is often represented

> visually and habitually (*cf.* Chapter 8, sections "Grey minimalism" and "Channeling the shaman"). *The portfolios by the protagonists of sonic labor as well as their public presentations in various media formats and media platforms represent in a majority of cases the concepts of sonic cleansing as demanded by the silencing dispositive as well as the moralist demand for self-improvement as articulated in the theology of sound as audiopietism.* The blankness and the radical reduction to white fields, thin lines, and regular, symmetrical, circular structures (Figures 8.1–8.8) implies an undistorted situation, far from any interrupting noises. It also implies the easy to handle and explicitly explicated structure of all sound designing activities; the design process must appear as logically structured and neatly articulated and controllable (Figures 8.11–8.15). The affective and caring labor in the work of sound designers is carefully hidden as it would disturb the ideal concept of a crystal clear and cleansed signal process. Only the very personal habits, the stylings and accessories, the hairdo and glasses or gadgets a sonic prosumers carries around and works with personally, might grant a glimpse into some of this welcome disturbances and detours while designing sound (Figures 8.16–8.20).

These four playfields of sonic labor, *location and apparatus, skills and habits, conflicts and heuristics, portfolio and presentations*, regulate and inform the individual activities in designing sound. The complexity and also the embodiment in individual work biographies, working styles, and also work goals could be supposedly growing from layer to layer. But as is documented in the chapters earlier in this book these potentially broad if not infinite options are rarely exhausted. Could there exist a way to perform sonic labor as a generative labor that is *not* turning against its very protagonists? Could there be a sonic labor that does not discipline the urges of ambition by the corporeal effects of exhaustion and despair? Could there be a form of *regenerative labor*?

Regenerative labor?

What would be the characteristics of regenerative labor that could provide an alternative to exploitative forms of sonic labor? The condition of sonic labor is—in accordance with similar forms of generative labor in various areas of design and the creative industries—determined by the way that sounds are drafted, crafted, and disseminated, implemented, and maintained. The working conditions of sonic labor, though, are far from providing an environment or the prerequisites that might promote the best work on sound possible for the prosumers involved. Sonic labor operates, understandably, under the limited means of a cultural climate. The material culture is harsh, the silencing dispositive is strong, and conflicts as well as unfounded demands from all sides force sound designers to develop coping strategies and to cultivate habits to go on, against all odds, with crafting sonic artifacts. Sound design as a profession is, hence, not independent from the latest fads in management theory—including the dominant imagery and ideals of conducting and guiding a design process, also including strategies of organizing a group of collaborators, and obviously also the common truisms of investment, monetization and profit in the broad field of *generative labor.* The generativity of a sonic workforce that is expected by its clients and its contractors seems effortless—or so it shall appear at least. As a resource, it might have no end and hence is prone to excessive exploitation and to exhaustion beyond any corporeal needs and articulations of limitations and doubts. Any *economy of sound*—as any *economy of the senses*—is one that is rooted in historical dispositives of trade, of governance, of relationships, and of affects regarding technology, sensory artifacts, and sonic experiences. It is not an economy that is experienced with the specific corporeal restrictions and manifestations that are a constantly underlying constraint in caring labor and only punctually entering now affective labor. Bluntly said: exploiting the generative resources in humanoid aliens is not in all aspects identical with exploiting the material resources as crude materials, growing plants or reproducing animals. It is a much more malleable, frail, and continuously self-reflective resource. But actually: it is no resource at all. It is a source. A source without which sound design or any other form of design would merely not exist.

Generative labor is the mode of labor by which design ideas and, in the field of sound design, relevant sonic affordances in sonic materialities are developed as well as concepts for sonic patterns to address sonic consumers. This form of labor is guided, as shown, by the silencing dispositive and often moralistically scrutinized or even harrassed by audiopietism. This framework forces the sonic prosumers, on the one hand, to pursue their highest ambitions regarding sound and, on the other hand, to accept the unsettling reality of increasing exhaustion in the course of a year of work or a life of work. This development appears almost without any alternative. It might seem like a natural given, a life of work structured by necessities, but: "It is worth recalling that what is currently called realistic was itself once 'impossible'" (Fisher 2009: 17). Consequentially, the contemporary situation of sonic and generative labor is not ultimate. It is malleable. A different form of sonic labor is not only thinkable, it is also possible, it can be performed in different ways, with different material enclosures and different performative manifestations. This mere potential of differing future developments, though, can be understood as a work assignment:

> emancipatory politics must always destroy the appearance of a "natural order," must reveal what is presented as necessary and inevitable to be a mere contingency, just as it must make what was previously deemed to be impossible seem attainable. (Fisher 2009: 17)

But how would it be possible, how could it be possible to alter this state of generative labor, its aspects of exploitation and control into another form of labor, a kind of *regenerative labor*? The first step would probably be, and there is a long list of literature at hand that intends to go this way, to transcend the employment situation as the only framework for any labor and also for sonic labor:

> To emphasize how labor is not equivalent to employment also means to acknowledge how important free affective and cultural labor is to the media industry, old and new. (Terranova 2000: 46)

This transcending of the employment framework of course implies much broader consequences. It leads to a dispensation of monetary revenue systems, of hierarchical control mechanisms, of the hard sequence of admission, presence, and documentation that binds you or me in almost all situations of our everyday lives. It also leads to a dispensation of various economically hardwired dispositives of control and of silencing that are in place to secure and to maintain this employment framework. Yet, they could be altered one day—and at present times there are various efforts made in politics and in political theory that intend to move all in this direction: for exaple, the *Post-Operaismo*-movement, the movement for an unconditional basic income, or the derivative communism. At this point in the book I do not intend to scrutinize and to evaluate these contemporary articulations and efforts made to transcend the employment framework. Some of these efforts appear well founded yet almost politically non-negotioable, other efforts though seem rather lofty, even not thought through on all societal consequences but have a good chance for being experimentally realized and maybe implemented as a sociopolitical standard in more wealthy nation states in the near future. In this section though, I intend to discuss a more fundamental turn in the concept of generative labor in paticular. I like to discuss the question of the imagined concept of regenerative labor: *does regenerative labor actually exist?*

From our personal experiences and the interviews and field research trips with sound designers we can say: *regenerative labor does exist.* It exists in zones and moments of joy and of pleasure in crafting an artifact with the maximum of space and time needed. It exists and it expands its energy and its subsequent effects into other activities, other forms of labor, into other moments experienced. It becomes possible in a situation of labor that exceeds the common functionalized reduction to producing monetary value at the workplace—and that contributes to the well-being, the development, the relations, and not least the material revenue among the members of a sonic workforce. Regenerative labor is a form of labor that generates, invents, models, crafts, and finalizes an artifact with no apparent or pressing limit in time schedules or in spatial limitations. It represents a zone of individually perceived freedom and of personal expression that is not necessarily an actual and observable, provable fact in a working situation. In almost all cases, though, this includes—unsurprisingly—a fixed and more than just sufficient salary. This remuneration is the main investment in the possibility of regenerative labor (which otherwise would remain only a feeble rhetoric, maybe a corny team building strategy). Yet, it does require a certain framework to constitute such a materially factual regenerative labor: a specimen of labor that foremost does not exhaust its protagonists—but empowers them and supports them in accumulating more energy, more individually desired fulfillment, and in general provides them with more access to all forms of participation in society. Such a regenerative quality of labor does to a large extent already exist in the working life of the earlier mentioned *hermeneutic class*—it does not yet exist though for the *exhaustion class* to which most sonic prosumers sadly belong (Lippe [1987] 2000: 295; *cf.* in this chapter the earlier section "Ambition and exhaustion").

A framework of regenerative labor would reinstate the respect and the individual freedom for sound designers and designers of all kinds. It would not force them to work as third- or fourth-rate contributors, doing largely disrespected yet absolutely necessary work. And, again, I do not speak here about the few star designers and celebrity creators in the field; I am speaking about the large, often invisible and seemingly endless personnel of the sonic workforce: a workforce that is pressed to work for less and less revenue and in many cases even to consider their work as a kind of showcase and public presentation that does not need any kind of proper and fitting honorarium. If regenerative labor would exist for a majority or even the totality of this sonic workforce, then its protagonists would receive a recognition that would resemble the recognition of an author, of an inventor, of an influential consultant. They would be cherished. As regenerative labor it would be given an open timeframe to complete the task; the individual sonic patterns and newly discovered sonic affordances would be met with respect and with interest. And individual sonic artifacts would not be seen as minor toys but as major statements. One cannot, obviously, reverse reality to such a dreamscape, for sure; but one can contribute from all sides—as client, as contractor, also as team member (and as researcher)—to build an environment that considers regenerative labor and all its aforementioned effects as the main motivation and goal. Regenerative labor is, therefore, also a part of conceiving and writing this section about regenerative labor in a writing environment and a moment of the day and the year that was not assigned by a time schedule or a workplan; but by my, the author's, individual urge to craft these arguments, these narrations, these chapters. In academic work, at least partially and punctually, regenerative work still does exist and can be performed—but maybe not for long. Regenerative labor is a concept of labor that is not exploiting and combusting the lives of humanoid aliens, but a sustainable and materially realistic form of labor regarding the bodies, the sensibilities, minds, and lives of those working in it. Instead of abolishing this specimen of labor as sticking to a nostalgia referring to old and outdated, preindustrial work environments, one could project regenerative labor as a far out but reachable utopia of the twenty-first century. Why not imagine regenerative labor as the first version of a specimen of labor that expands our notions of monetization, of capitalization, of exploitation and surplus value? Maybe regenerative labor could be a starting point to move present societies of employees and consumer citizens into collectives of sensory personae encountering each other and collaborating as humanoid aliens? Regenerating a utopia; in times of growing dystopia.

15

The Panacoustic Society

A sensology of surveillance

Ubiquitous domestication

Sound design in the twenty-first century is contributing through a process of ubiquitous domestication of consumption and its dispositives to the forces of authoritarianism that promote a sensology of surveillance, on which the panacoustic societies of the near future are currently being founded: this is the third major result from our fieldwork at sound design projects of differing scales and of varying applications as well as our critical analysis of the history, the design approaches, and the listening practices of sound design. Ubiquity is undoubtedly one of the key characteristics of contemporary technological inventions—as well as an assumed trajectory of technology in the nearest future. This ubiquity, though, also results understandably in a thoroughly ubiquitous character of the effects of these new technologies. This also translates into a whole new lifestyle attached to a strand of economic formations, of sensological dispositives and practices, as well as a sound culture that result in a new social constellation. In this chapter, we would like to unfold this concept of society under the title of a *panacoustic society* (Schoon 2012) or the *panacoustic supplement to the panopticon* (Szendy 2016: 16–23) mainly driven by a form of *surveillance capitalism* (Zuboff 2015, 2017) and its related legitimization discourse sometimes camouflaged as altruist *data-philanthropy*. Starting with the ubiquity of technology the aspects of a dominant silencing dispositive (Chapter 13) and the related economy of sonic labor (Chapter 14) have already been explored in this fourth and final part of this book. Both of these fundamental constituents of sound design as a profession in the twenty-first century are embedded in the dominant changes in the culture of technology and its effects on society and governance as already pointed out in reference to dispositives of control. Yet, the impact of social and political changes in society run much deeper than just a certain redirection of impact or of focus in research: it actually reorganizes the whole of the design practice, the goal of generative labor, and its instrumentalization in society. This instrumentalization of sound design starts only with the aforementioned aspects of a dispositive of control.

The ubiquity of technology affects the ubiquity of sound. As thoroughly analyzed and convincingly argued by Anahid Kassabian (Kassabian 2013), the dispositives of musical consumption and musical production have gradually transformed from the earlier, largely class-centered and pedagogically as well as militarily structured dispositive of the conservatoire and the concert hall, to a dispositive of ubiquitous technological distribution of and mass cultural participation in the production of, the discourse around, and the evaluation of sound and music. More and more listeners, musicians, and *prosumers* of music are inhabiting a radically transformed world. This transformation, though centered around music consumption and production, also affects and is symptomatic in the transformations of sound design. Actually, the ubiquity of listening and music is almost unthinkable without the ubiquity of sound and its design: as a ubiquitous phenomenon, listening and producing music is strongly attached to the *functional listening* practices related to sound design (as explicated in Chapter 13, section "Functional listening"). One could even dare to say that the hegemonic impact that once was radiated by music now has already gradually shifted more and more into sound design: whereas music was widely regarded as a major cultural force and layer in recent decades in popular culture, this might have switched recently to sound design. We might elaborate on this issue in another future study of another research project.

At this point of our exploration on the social impact of a transforming governance around sound design, it is necessary to unfold the quality of this ubiquity. As stated earlier (in Chapter 10) it is necessary to acknowledge that: *perceiving and living with sound is a performative act with a situative and a sensory character that constitutes one of the technologies of the self.* This leads us then to the impact of domestication in listening and sounding, especially following the approach by Roger Silverstone, and to the question: *does the domestication of new sound designs also involve, quite literally, a taming of the wild and a cultivation of the tame?* (Especially in the section "Domesticating sounds" of Chapter 10.) At this point now, answering this rather excessively open question seems at least pressing and maybe at least tentatively possible. Whereas earlier we explored the corporeal and situated aspects of domestication in contrast to earlier approaches from semiotics, the general impact of domestication is here our issue: *How does a domestication of sound and the senses transform a society also undergoing a transformation toward the ubiquity of access to information technology and especially to sound productions?* What is the effect of this *Ubiquitous Domestication*?

One major effect includes, quite obviously a crucial leap from quantity to quality: if a larger amount of technology is ubiquitously online, connected, and in constant exchange then a constant process of reciprocal domestication will also be taking place—an incessant "taming of the wild and a cultivation of the tame" (Silverstone 1994: 11). Technological inventions no longer enter one's everyday life only every few years or decades, or once in a lifetime, now such encounters happen, in these days of the late 2010s, once a week or once a season if not every single day. It is hence an ongoing if not unspoken obligation of a true consumer citizen to engage in these processes of domestication, involuntarily or indeed with their full heart. Therefore, it is even more an obligation of the sonic consumer and the sonic prosumer—the sound designer. This process of an ongoing, an incessant domestication in both directions, a taming of new wild technological commodities as well as a cultivation of you and me, the ones trying to tame and to domesticate certain commodities, this process is now dominating individual lives. Each consumer citizen might decide individually on how intensely she or he will engage in this process of domestication, but actually there is no way out. Evading this process at all is simply no option you or I are even allowed to select. Choosing this option resembles more of a radical opting out of contemporary society and often also contemporary habits, ways of communicating, and partaking in the everyday life of a majority of the other consumer citizens living around you. It is clear, though, that the problematic aspect in this equation is not so much the ongoing process of domestication in ubiquity but the role of the citizen consumer and the expectations directed toward this actor.

Going back to the framing analysis of this book regarding the role of listeners, designers, consumers, users of sound design and music productions (throughout the Part One: "Why That Sound?," especially Chapter 3) it becomes clear that the shift in the role of the citizen—as now being inextricably tied to the role of the consumer—has transformed every single meaningful social context. It paved the way toward putting citizens almost exclusively on the receiving end of social developments and political decisions. It results in a discouraging and a disengaging, a form of dominant depressive inner emigration in everyday life: *if there actually is no society anymore, why would I be required to partake or even to engage personally in this thing that apparently does not exist anyhow?* The neoliberal project of an actually invasively pervasive and perverted hypercapitalization and hypercommodification, the actual prioritization of profit and capitalization on all levels of society, of private existence, and of your and my intimate life would therefore have reached its most impactful goal.

If indeed this would be the case, then one would necessarily have to argue that the process of technological domestication and its ubiquitous distribution indeed contributed mainly and foremost to the domestication of its consumers, its users, its prosumers who assume they would only domesticate their new consumer electronics gadget. This is the double bind that apparently seems almost impossible to overcome or to dismantle: without the establishing of consumer citizens in recent decades as the major model for citizenship it would also not have been possible to implement this large-scale process of ubiquitous domestication; and in reverse, this ongoing process solidifies the model of you and I mainly being reduced to the rather obediently functional role of consumer citizens. It seems now to be a self-supporting structure of two interdependent dispositives. Sonic prosumers love being consumer citizens and domesticating new audio technology commodities. Sonic labor, ubiquitous domestication, and the silencing dispositive are tied to each other and supporting each other. Right now, at least, it does not seem as if this mutually supporting structure is being easily dismantled and disassembled in the near future; though this might nevertheless still on the erratic course of sociopolitical serendipity. Still, there might occur a certain, minor chance of discovering certain cracks, instabilities, intrinsic inconsistencies that already show today—and this might, hopefully, result in selected, powerful reverse actions or massive counter-movements. But this, again, is only part of utopian imaginations—not part of dystopic realities.

Forces of authoritarianism

What are the forces of authoritarianism that promote a sensology of surveillance in panacoustic societies? The condition of an ongoing and ubiquitous domestication of and by consumer citizens is not a simple one to dissect. Especially as the development, the invention, the implementation of, and the propaganda for further new artifacts of technology is progressively moving on as we speak, as I write, as you read these words. Irresistibly, this storm drives you and me into some yet unknown situation to which our backs might still be turned right now—while the rubble-heap before us grows sky-high. "That which we call progress, is *this* storm" (Benjamin [1940] 2005; emphasis in the original). This continuous technocultural push, however, does not seem too arbitrary any longer in the late 2010s. Its trajectories, ramifications, and radiations seem more and more predictable in times when "the consumer-oriented liberalism of Bill Gates and Steve Jobs gave way to the *technological authoritarianism* of Elon Musk and Peter Thiel" (Haider 2017; my emphasis). For a cultural theory of sonic labor and sound design culture such concepts as *surveillance capitalism* (Zuboff 2015) and *panacoustic society* (Schoon 2012, Szendy 2016) provide a necessary starting point to scrutinize contemporary developments in politics and culture: are we experiencing a legitimizing and a normalizing of practices, technologies, of research and development that promote authoritarian policies? It seems necessary to distinguish here various degrees of critical, uncritical, affirmative, strategic, and tragically irresponsible and naïve efforts. For this distinction I would like to propose four categories already introduced in this book as a new conceptual framework to understand current developments toward authoritarianism: *consumption, domestication, dispositive*, and *ubiquity*. These categories might contribute to explicate these current transformations; transformations that effect and abuse pervasively the culture of sonic labor and of sound design.

These four categories, introduced partially in Part One of this book—"Why That Sound?"—and further developed and explicated in this fourth and final part of the book have been used here to outline the actual cultural practices of crafting and of living with sound design. All four are—as will become apparent in the next paragraphs—also the main constituents for a development toward a pervasive use of data collection, of data mining, and of the use of data for surveillance purposes. *Consumption, domestication, dispositive*, and *ubiquity* are forces that characterize consumer products that seem at first to be just a new appendix connected to the consumer—but in the end they actually turn the consumer into a domesticized appendix of this new product. Be it a hardware gadget or a software app, a network you are permanently logged into or a hobby, an amateur fad, they draw you into this forcefield of widespread connectivity, of prolific tracking and profiling and pattern recognition. You and I, we are being consumed and domesticized in this ubiquitous dispositive—and we like it. This ongoing process of domesticating through consumption has turned into an unstoppable and apparently often an uncontrollable one; even more so, as ubiquity is its most prominent characteristic. However, to understand the impact of this process the role of the overarching dispositive has to be scrutinized, that is stabilized and confirmed by these activities. It is the *silencing dispositive* of cleansing, filtering, and purifying that the aforementioned other three forces actively nurture, confirm, solidify, and even expand. It is through the silencing dispositive that you and I are being consumed, that we are being domesticated, that our domesticated existence of consumption is being expanded into a nearly global ubiquity: this dispositive of authoritarianism implies and promotes any useful form of violent cleansing, of exclusion, of domination and erasure that a silencing dispositive derived from signal processing theories might offer. What might seem indeed necessary in all minor and situated examples for focusing and reducing disturbances, for stabilizing a core interest or a firm ground, turns toxic and lethal when expanded beyond the appropriate realm of one's individual life and specific experience. The generalization of idiosyncratic preferences again gives birth to totalitarian monsters of annihilating desires.

The intrinsic interpenetrations between the four forces now deserve a closer look. *Consumption* and *domestication* can be considered as representing mutually fundamental conditions: they evolve out of each other as major categories of everyday life, of consumer citizenship, and of mere existence in consumer cultures and technoculturally driven societies—at least since the nineteenth century with the inception of Western industrialization. The two other categories though, *dispositive* and *ubiquity*, are not necessarily foundational to everyday life as they are far more relying on rather arbitrary developments of recent decades—only tentatively beginning in the second half of the twentieth century with the dissemination of Westernized mass media and its global networks. Briefly said: *consumption and domestication dominate societies, on the one hand, without necessarily implying the ubiquitous distribution of sensory dispositives—these ubiquitous dispositives, on the other hand, are strictly unthinkable without relying on the grounding practices of domestication and consumption*. Between these four powerful and radiating nuclei of *consumption, dispositive, domestication*, and *ubiquity* the known forms of surveillance emerge as the culturally and societally structuring forces one can observe in the twenty-first century, expanding and colonizing all areas known to humanoid aliens. In conclusion, these four categories represent the generative forces that conceive contemporary forms of authoritarianism.

The further *sensological* (Perniola 1991) effects of such a world being colonized by expanding surveillance will be discussed in the following section of this chapter. In this chapter though we will focus more on the immediate effects emerging out of this forcefield generating ubiquitous and authoritarian control—and how this model can be used to distinguish the aforementioned degrees of critical, uncritical, affirmative, strategic, and tragically irresponsible and naïve efforts toward authoritarianism. As soon as this forcefield of consumption, domestication, dispositive, and ubiquity is installed in a culture, the social relations, professional activities, the intimate experiences, personal expressions, and the political constellations will transform and shift accordingly. The individual forces in this fourfold field might be easy to overlook; however, over time it has been shown that precisely these forces do not remain chaste. They take effect and they shape the executive, the jurisdictive, and the legislative power of the modern nation state more and more. Consuming industry products demands more and more industrialization of consumption, a habit of domestication demands more and more habitualized domestication, solidified dispositives require more and more solidifying of dispositives, and an expanding ubiquity fosters a further and further expansion of ubiquity. Self-fulfilling prophecies of progress, surplus, excess, expansion, and annihilation of all other are turned into reality in the dimensions of planetary annihilation and thanaticism (Wark 2014) in the *Capitalocene* (Moore 2014a, 2014b).

So, how are these four forces of consumption, domestication, dispositive, and ubiquity exactly fostering a culture of authoritarianism, in individual practices and in general development? These four forces individually might not necessarily enforce a movement in the direction of authoritarianism: consumption is by some advocates of liberalized market economies considered a fundamental democratic right; domestication can be regarded as a highly idiosyncratic practice; dispositives can be considered the results of a dominant culture of entrepreneurship; and ubiquity is a quite obvious result of a highly competitive and only punctually collaborative market economy. However, if these four forces are connected into one seamlessly and incessantly operating continuum, then this continuum seems to generate foremost not products, consumers, market opportunities, or revenues but a consistent authoritarian system. This system is being generated through the parallel presence, activity, and influence of all four forces. From the angle of the desiring practice of *consumption* a force is encouraging contemporary citizens toward an accelerating excess acquisition of commodities, products, new gadgets, and newly presented and packaged goods. At the same time a force from the corresponding angle of the habit of *domestication* is encouraging citizens toward an equally accelerating process of domesticating and appropriating these new commodities and products as if they truly were in the possession of the consumer citizen. A force from the opposite angle of familiar *dispositives* also is at the same time positioning consumer citizens into its framework as a basic necessity for the first two processes of consumption and domestication to continue. And, finally, from the angle of *ubiquitous* presence a force reaches out to citizens to never stop, reflect, hinder or even reverse this fourfold process of fusioning forces as this would have catastrophic consequences for the lives and the future of all citizens involved. This forcefield therefore stabilizes itself and it does so through the promise of further continuation of the current state of a relative wealth, a relative security, and a relative future perspective. There is no perspective, however, outside this current state of affairs: the quite authoritarian request to support its established power structure does mainly serve its own continuation. *Consumption, domestication, dispositive*, and *ubiquity* have ceased to—and maybe they never actually did—contribute any further development to the current state. Apparently, they are mainly supporting an excessive prolongation of the current status quo: one of the most horrifying dystopias, effectively.

The sensology of surveillance

How can a sensology of surveillance be described on which panacoustic societies are being founded? The ongoing normalization and legitimization of the four forces of authoritarianism, *consumption, domestication, dispositive*, and *ubiquity*, is no recent development. This process began decades ago and underwent a further stabilization, an emphasis, and an expansion in quantity. Then, in the early twenty-first century, with the quite welcome advent of the *Global War On Terror*—yet another *invented tradition* (Hobsbawm and Ranger 1983)—the domestication of various specimens of surveillance technology reached new dimensions. The quality of the normalization process changed and this new quality of normalization deeply affected the sensorium of us humanoid aliens turned into consumer citizens. *Sonic consumers of the early twenty-first century got the impression they had next to no choice than to contribute by all their activities to a growingly ubiquitous surveillance dispositive—simply by consuming and appropriating their new consumer products.* This expansion of surveillance manifests itself not solely in its quantity and ubiquity—these are just its most obvious characteristics; foremost this expansion manifests itself in a *sensory* implementation that eradicates remnants and traces of any surprise, irritation, any stronger resistance against this *ubiquitously domesticized consumer surveillance dispositive*. This expansion qualifies truly—following Mario Perniola—as an implementation of a new *sensology* with its main goal to govern present and future societies and cultures by authoritarian exclusion of critique and opposition (Perniola 1991). Perniola defines sensologies by their shift from a scripture dominated ideology to mainly sensory and embodied forms of ideology (Perniola 1991): sensologies are ideologies of, in, and through the senses. They are hegemonic in all situations and areas where someone tends to question the actual existence of a sensological ideology: *Well, don't be a spoilsport, don't overthink this, can't you just enjoy this? It's just fun, right? It's sexy, isn't it? And it gives you a great feeling, no? Don't be bothered with that theory crap.* This shift from scripture to sensorium can be observed in a number of apparatuses and practices that contemporary technological developments recommend their citizens to consume and to domesticize. In the following three examples I would like to analyze how such a sensology of surveillance is being normalized and implemented through three examples of current *sensotechnoculture*: the *baby video monitoring*, the *Quantified Self* movement, and the *voice assistants* such as Siri or Alexa.

Baby monitors have been quite common since the 1960s (Mihm 2016). Yet network and broadband technology, wireless communication, and the growing ubiquity of video recording has transformed and expanded the once simple and timid live audio transmission from the cradle where a baby sleeps. In recent years the baby monitor technology has tried more and more to mimic the familiar video conversations on mobile devices. Recent products such as the Summer Infant's *Baby Touch WiFi Video Monitor*, the *Angelcare 1200 Video Movement*, or Samsung's *SmartCam HD Pro monitor* make it possible for a larger number of friends or relatives to watch this stream, to interact with each other, to record, and to store their favorite recordings. This technology, hence, is perfect for the normalization of the surveillance dispositive: the parents learn to appreciate the benefits and the joy of permanent audiovisual surveillance, the kids learn and normalize the permanent presence if not dominance of a recording and transmitting device; and in growing older both might have a really hard time in objecting against any sort of ubiquitous surveillance in other areas of life. *We have known this since we were kids, we have nothing to hide, it is after all just a way of life ...* What sounds dystopic here and now, might be falsified by future history—and, frankly, I sincerely hope so. Yet, I expect more that these critical notes on surveillance technology will only be read then as ridiculously antiquated concerns by someone who simply could not truly anticipate the future lifestyle with all its pleasures and excitements. This impression surely is right: once a sensology of surveillance is implemented, it might still be possible to imagine life without it—but it becomes harder and harder to *desire* this earlier but poorer life (and not just as a holiday experience of detox or disconnect or as means of distinction). The parental desire is exploited here in a classic example for the perfidious functioning of a sensology. One crucial step further in the implementation and domestication of surveillance technologies is represented in the extreme practices of the *Quantified Self (QS)* movement that has evolved since 2007. This movement grew out of the hacker and cyborg community and gained traction as Gary Wolf and Kevin Kelly, writers for the magazine *WIRED*, officially founded *QS* and, subsequently, this movement assimilated all sorts of apparatuses to measure, to record, to compare, and to model any individual's material and physical behavior. The protagonists apparently regarded this practice as liberating, as a newly gained autonomy if not a rebellion against corporate measuring gadgets like the ubiquitous smartwatches for athletes. However, also in this case the first liberating effort of a new individualized and personalized technology quickly is successfully assimilated and soon thereafter usurped, derailed, privatized, and irrevocably commodified by one or various corporations. Well-documented earlier examples for instance of terrestrial radio, of tape and video recording, and more recently the internet-related examples of weblogs, of podcasting, and of free Wi-Fi can serve as a sufficient sample here. Quantifying one's individual activities throughout every day, therefore, has become the most obvious way of domesticating pervasive surveillance under the more attractive moniker of individual self-exploration. This latter goal might still be achieved individually—yet the abuse and monetization of exactly this personalized and individualized practice devalues this goal as just another clever marketing strategy: an exemplary sensology to exploit personal if not intimate desires in order to capitalize them (Mau 2017). The third and final example for contemporary sensologies of surveillance in the process of domestication are the well-known *intelligent virtual assistants (IVA's)* such as Siri or Alexa (Hoy 2018). The long history of artificial voices, personal assistants, and a guiding agency in buildings or as the sonification of a map started out in (science) fiction—and the fantasy of a robotic (often female) slave, servant, or assistant has transgressed into a contemporary startup- and venture capital-economy, almost unaltered. It might be trivial to diagnose in this case the normalization of the surveillance dispositive; yet with the ubiquitous domestication of such assistants the domestication of surveillance in general achieves a level of laconic pervasiveness that both aforementioned examples—the *baby video monitor* or the *self-quantification movement*—never achieved. The consumption of these artifacts and vocal service personae is not only plausible or even attractive, it appears as a logical if not necessary consequence of contemporary technological development: a consequence that seems highly desirable for everyone's everyday life (Power 2014). This is the constellation of desire that is exploited by developing, offering, and selling such assistants. Exactly this exploitation of contemporary desires and needs to a maximum in order to capitalize them is the main characteristic of a sensology. Voice assistants of any kind are hence a dispositive that is almost irrefutably entering contemporary homes, workplaces, and the public sphere—and with this unstoppable expansion the highly addictive consumption of all services related to these assistants is not only introduced but forced upon every consumer citizen.

This whole framework of a sensology of surveillance though mainly takes up again the desires and the utopian goals that have been originally developed and criticized in the framework of the *Californian Ideology* (Barbrook and Cameron 1996), then mutated and migrated into the discourse and its related practices of the *Creative Class* (Florida 2002), and finally taken up into what McKenzie Wark recently coined as a *Vectorialist Class* (Wark 2012: 137–148). The *Californian Ideology* (Barbrook and Cameron 1996), very present already in the 1990s, promotes a kind of solutionism that assumes that each and every desire, need, any problem or trouble a humanoid alien could encounter could perfectly be met and solved and actually instrumentalized by a technological invention or application that would make life easier. Factually, this urge and this desire has been executed and implemented in contemporary startup-culture since the 2000s and affirmatively promoted as a concept of the *Creative Class* (Florida 2002) that allows for a professional life that still is in constant touch with everyday problems and with the individual desires for artistic expression, for independent labor not framed by employment, and for, what we called earlier, *regenerative labor*. Both earlier concepts—*Californian Ideology* and *Creative Class*—then merged recently in the concept of the *Vectorialist Class* (Wark 2012): this class of high ranked and top-earning executives, inventors, directors, and consulting personnel indulges in all of the recent gains and benefits in order to execute power over all those underlings not capable, willing, or educated to operate on the metalevel of installing and orchestrating a vectorialist platform that is guidance, lawmaker, and beneficiary of this area of life, its everyday practices and all related businesses in this area at the same time. With this last step the outline of contemporary authoritarianism is phrased in a nutshell.

In the course of developing lifestyles, cultural practices, and business models from the framework of Californian Ideology—through the concept of the Creative Class—into the one of the Vectorialist Class, the sensology of surveillance evolved with it. The three aforementioned examples—of *baby video monitoring*, the *Quantified Self* movement, and *voice assistants*—taken together represent herein the cultural process of implementing surveillance as a major sensology in society and culture: these three cultural artifacts serve as carriers for sensory ideologies. The domestication of the sensology of surveillance proceeds in it mainly by a ubiquitous dissemination that establishes an almost irresistable dispositive for consumption. As soon as this dispositive is established, as soon as the ubiquity of this dispositive makes it impossible to imagine even mere existence without these commodifications, as soon as consumer citizens are unable even to question their longing for the consumption of such commodifications, then the sensology of surveillance is fully installed and it will seem unremarkable and invisible for those living in it. You might then perhaps no longer be capable of understanding these lines, this section, or this whole book. You might wonder, why this old bloke, born in the late twentieth century, rambles on about such bare necessities. You might read these lines only as an ancient and ridiculously freaked out, quirky rant. The present dystopia appears then as the most laconic normal life; the same as earlier dystopias have been normalized as well in the recent course of history. The realization of any dystopia is as invisible and seemingly negligible as the realization of any utopia.

The panacoustic society

What are the constituents, the practices, and the ideological beliefs that characterize a panacoustic society? The new world emerging out of a sensology of surveillance is already very much tangible these days. But at the same time, it seems hard to imagine how life will be in this new world—without the slightest doubt about it. Still, there are symptoms of what this new world will feel like, sensorially:

> Over the past thirty years, capitalist realism has successfully installed a "business ontology" in which it is simply obvious that everything in society, including healthcare and education, should be run as a business. (Fisher 2009: 17)

This totalizing sensology of surveillance, founded on a supposedly infallible and unquestionable *business ontology*, constitutes, if you will, the fourth—and maybe final—step after the Californian Ideology, the Creative Class, and the Vectorialist Class. It provides a closure of this process that installs effectively a new social order. This new social order, which emerges out of various traditions from earlier societal models, is centered around everyday practices, technological means, and political discourses around surveillance. As soon as the forcefield of authoritarianism consisting of *consumption, domestication, dispositive*, and *ubiquity* is implemented in a society it reaches a state of normalized *hypercommodification* that resembles a surveillance state in all its relevant aspects—aside from an explicit political and constitutional commitment to it. The Swiss media studies and cultural studies scholar Andi Schoon and the French philosopher Peter Szendy both applied the term *Panacousticon* to this kind of a pervasive social constellation of surveillance mainly generated through sonic technologies. Szendy and Schoon use this word for an auditory interpretation of the originally visual and architectural surveillance structure established by Jeremy Bentham's notorious prison dispositive from 1791; a model that was scrutinized and brought back almost two hundred years later into the common cultural memory by Michel Foucault's *Discipline and Punish: The Birth of the Prison* (1975). The main point in Foucault's interpretation was—in Schoon's words:

> In Bentham's model only one person is accommodated per cell, for which the guardian disappears in the backlight. The prisoner can therefore never tell whether the tower is occupied at all. But, according to Bentham's innovative idea, he will always behave as if the tower was occupied. As a result, the prisoner monitors himself. His model is extremely rational, it individualizes the prisoner and aims to make him useful again for society. He who learns to behave will, after a certain time, be left to freedom and thus to himself. (Schoon 2012: 296f.; translated by Holger Schulze)[1]

This *Panopticism* can be considered a general model of society and of a mainly visual historical sensology. In the words of Michel Foucault:

> On the whole, therefore, one can speak of the formation of a disciplinary society in this movement that stretches from the enclosed disciplines, a sort of social "quarantine," to an indefinitely generalizable mechanism of "panopticism." Not because the disciplinary modality of power has replaced all the others; but because it has infiltrated the others, sometimes undermining them, but serving as an intermediary between them, linking them

[1]"In Benthams Modell ist pro Zelle nur eine Person untergebracht, für die der Wächter im Gegenlicht verschwindet. Der Inhaftierte kann also nicht erkennen, ob der Turm überhaupt besetzt ist. Aber, so Benthams innovative Idee, er wird sich stets so verhalten, als sei der Turm besetzt. Im Ergebnis überwacht der Gefangene sich selbst. Sein Modell ist äußerst rational, es individualisiert den Gefangenen und zielt darauf ab, ihn wieder fürs Gemeinwohl verwertbar zu machen. Wer lernt, sich zu benehmen, wird nach einer gewissen Zeit in die Freiheit und damit sich selbst überlassen" (Schoon 2012: 296f.).

> together, extending them and above all making it possible to bring the effects of power to the most minute and distant elements. It assures an infinitesimal distribution of the power relations. (Foucault 1995: 216)

Szendy and Schoon transfer this visual prison dispositive of the eighteenth century and its *infinitesimal distribution of the power relations* into the sonic surveillance society of the early twenty-first century (*cf.* Szendy 2018). Also, by extracting its original sonic traces:

> this visual architecture, this all-seeing machine, goes hand in hand with the possibility (that Bentham envisages without truly developing) of an apparatus that we could call panacoustic. ... What Bentham imagines here as an addition or a supplementary accessory to his Panopticon is a simultaneously pandirectional and selective megaphone of sorts: a Panacousticon that facilitates communication and transmission between observer and observed in the context of efficiently organized labor. (Szendy 2016: 19f.)

The roots of this *Panacousticon* in *a simultaneously pandirectional and selective megaphone of sorts* could then even be traced back to Athansius Kircher's famous image from the seventeenth century of a house in which gigantic ear trumpets are built in to intercept all audible activity in the inner courtyard of the house. Also this genealogy would follow coherently the architectural dispositive as interpreted by Foucault in Bentham. But on top, Schoon traces the Panacousticon back to the invention of the *stethoscope* (Rice 2012) by René Laënnec in 1816 with its goal to use actual *non-speech audio*—the sounds of the human body—to retrieve and to convey information. This goal, though, of the inventors of this seemingly innocent technical tool is identical with the alleged definition of *sonification* and any form of auditory display and functional sound design in general: "Sonification is the use of nonspeech audio to convey information" (Kramer et al. 1997).

This is the exact pivotal point of transition into a surveillance society: as soon as forms of auscultation and sonification are sold as ready-made tools for *consumption*, which establish by its *domestication* a whole *dispositive* with an *ubiquitous* character, the sensology of surveillance by sound and by all other senses is not only in principle theoretically possible, but it is introduced and it is suggested as a reasonable and sensible technology for governing and for an *infinitesimal distribution of the power relations. Sound design, in this interpretation, resembles a functionalization and an informationalization of the audible; as such it opens up a two-way road of signaling and audio monitoring, of militarized sonic signals and of militarized surveillance of any sonic traces, of sonic cleansing and purifying from unwantend sounds, from noise through efficiently organized sonic labor.* The silencing dispositive becomes state belief. Schoon writes:

> Sound designers report on the optimized feedback from the machine, logistics experts hear the delivery bottleneck clearly, psychologists introduce new insights into the reception behavior, also chemists and astronomers are now listening more clearly than in the previous year. (Schoon 2012: 298; translated by Holger Schulze)[2]

[2]"Sounddesigner berichten vom optimierten Feedback der Maschine, Logistik-Experten hören den Lieferengpass inzwischen deutlich, Psychologen stellen neue Einsichten in das Rezeptionsverhalten vor, auch Chemiker und Astronomen lauschen jetzt klarer als im Vorjahr" (Schoon 2012: 298).

This operationalization and this semiotization of sounds and their effects establishes all four major forces of authoritarianism; forces that are needed to maintain and to expand a surveillance apparatus. This apparatus is characterized, as Szendy writes, by a sort of:

> panacoustic hypertension that does not want to lose anything at all from the whole, the detail, never immediately given. (Szendy 2016: 115)

Hence, sound design is, voluntarily or involuntarily, a main supporter and a contributor to the sensology of surveillance that is taking over more and more areas of everyday life in the early twenty-first century.

> The insufflations within the Panacousticon are even more difficult to identify than the visual axes of an ideal prison architecture. In short, the "auditory turn" has an affirmative downside, which should not be left unheard in a sonification discourse that is all oriented toward progress. (Schoon 2012: 299; translated by Holger Schulze)[3]

A sonic society is a *Panacoustic Society*—under the conditions of a pervasive, sensology of surveillance. This massive modernization turns the hitherto impossible into the now possible and in reverse:

> It is worth recalling that what is currently called realistic was itself once "impossible": the slew of privatizations that took place since the 1980s would have been unthinkable only a decade earlier, and the current political-economic landscape (with unions in abeyance, utilities and railways denationalized) could scarcely have been imagined in 1975. Conversely, what was once eminently possible is now deemed unrealistic. "Modernization," Badiou bitterly observes, "is the name for a strict and servile definition of the possible. These "reforms" invariably aim at making impossible what used to be practicable (for the largest number), and making profitable (for the dominant oligarchy) what did not used to be so. (Fisher 2009: 17)

The hitherto impossible, permanent active apparatuses of a pervasive control of sonic consumers and prosumers, using the ubiquitous dispositive and the domestication of consumption, they achieve altogether a silencing and an exploitation of humanoid aliens in applying audiopietism as its governing sensology.

[3]"Die Einflüsterungen innerhalb des Panakustikons sind sogar noch schwerer zu identifizieren als die Sichtachsen einer idealen Gefängnisarchitektur. Kurz: Der 'auditory turn' hat eine affirmative Kehrseite, die auch im fortschrittsorientierten Sonifikationsdiskurs nicht ungehört bleiben sollte" (Schoon 2012: 299).

16

The Sonic Capital

An anthropology of sound design

The situation of sound

Sound design in the twenty-first century is endangered by solidified emanations of sonic capital. Its practices of sonic labor are in dire need of being regenerated and remediated by the nanopolitics of experiential listening and generative sounding: this is the fourth result and concluding recommendation from our fieldwork at sound design projects of differing scales and of varying applications as well as our critical analysis of the history, the design approaches, and the listening practices of sound design. This is now the situation. In front of me I can look at a screen with smaller or larger loudspeakers attached to it; I can hear the high frequency of the machine working, I can hear the sonic signals of this networked and mediatized computer environment—in my laptop, my VR-headseat, my AR-glasses, my smartphone, or my tablet. I am outside, I watch other humanoids walking by, I await other citizens approaching me, for conversation, for collaboration, for serendipity. My torso and my limbs are warm, I can sense the tension in my nerves and muscles, I swallow some saliva, I breathe calm but with a certain joy that is full of positive expectations. In this situation, I am present or distracted, I observe and listen or I take action, I articulate or I reflect, I sense. In this situation, I encounter crafted sound events incessantly, routinely and in a frequency and a damped quality that I all too often miss them as such. I do not hear them anymore—yet I sense them. The sounds, digitally transmitted or generated, they are in my body now, in my general feeling of who I am, where I am situated right now, and what seems to be expected from me or what seems to be unthinkable to do right here. These sounds, almost inextricably amalgamated with all the technological apparatuses, the furniture, the decoration, the architecture, the visual guidance in this place as well as the various humanoid and nonhumanoid aliens present, moving, and performing in this place, they constitute this moment. This sensory, this material sense of the situation is *the substance of this situation.* The substance is, hence, not a structural, static, or a formulaic one. On the contrary, it consists exactly of the highly dynamic and relational trajectories and their individually intrusive qualities of all the sensory effects. It is the substance of my or your individual experience, the immanence in all its situated details, in all its remaining and still continuing effects from earlier encounters, pivotal biographical instances, and the hangover of historical events. Sound design takes place and time in exactly this substance: sound design operates in all of these substances.

The situation of sound design—in all its *prosumptive qualities*: in production and in consumption—is a personal and an intimate one, to the same extent as it is a social and a public one. The substance of a sonic experience is to be found precisely in this limited situation of sound and listening—following at least an anthropology of sound as an interpretation according to sonic materialism (Cox 2011): limited in time and in space, embedded in cultural and historical contexts, in material and sensory constellations and intensities as well as all their continuities and discontinuities. The situation of designing sound is, hence, implemented, derived from, and effecting this very situation where sound can propagate, resonate, can be reflected and be transformed. The freelancer's studio and laptop or the sound design department's basement with small recording booths and an intimidating big mixing desk is the laboratory, but the sonic artifacts crafted therein are *renaturated* and even *rewilded* in closed office spaces, on plazas, in transport vehicles that are far from soundproof or acoustically optimized. Exactly this wild outside of everyday life is the major joy as well as the major problem of any effort to design, to create, and to craft sound events with a more or less functional character.

As living with, perceiving, and intervening into sound are performative acts within substantially sensory constellations, these acts are generative of the substance of these constellations, and at the same time, these acts are themselves grounded in, and influenced and attracted by those very sounds present in a specific situation. The substance thus changes as one makes meticulously planned efforts to mould this very substance. As every sound designer is operating in such a moving, a plastic and shapeable, an incessantly resounding situation, this resembles truly a minuscule surgical operation in the tiniest and most interconnected organs of a person while she or he is living, moving, working. Obviously, one can test sound designs, in the public sphere and in their actual usage situations, but also workshop- and studio-situations. The lives of all the team members, clients, designers, and agency contacts are never disconnected from contemporary sound design. The sound design workstations themselves, the mobile communication devices, public transport, self-driving or not self-driving cars, all these apparatuses that we believe to be necessary for a fulfilling everyday life in highly networked, mediatized, and globalized societies and urbanized agglomerations of the early twenty-first century, operate while signalling their operating statuses with crafted sound emissions. These functional sounds enclose and frame our present situation. They shape a highly reactive, excessively complex, and constantly changing environment for any sound design. The everyday life practices with sound are seemingly always quicker in changing than any new sound designer is in developing new sound designs appropriate to this new situation of listening.

The sonic dominance (Henriques 2011) of these sonic signals of all sorts makes it almost impossible *not* to be affected by them. I am enveloped in beeps and buzzes, songs and jingles, announcing pings, confirming clonks, and rejecting plonks. These sonic signals contribute to the audible sensology of the present and they manifest the main societal and economical procedures, they engrave them in your or my sensory corpus. Whereas Henriques originally refers with the concept of sonic dominance mostly to bass frequencies, there is also a different dominance of mostly high-pitched functional sounds: this dominance resonates not mainly with one's guts and bones as bass sounds do but with the nervous system, with the urge to interprete and to act, to decipher and to respond. This is a situation of sensory intensity combined with a sonic agency. As a sonic consumer, there is almost no moment of my life in one of the urbanized agglomerations I am working and living in, I am experiencing and enjoying, in which functional sounds are not present. My functional listening activity is incessantly addressed and a wide variety of sonic patterns is transmitted into my individual, experiential situation of acting, perceiving, sensing, performing, and interacting. I am part, voluntarily or involuntarily, an integral part of the established economy of sound. This economy demands my reactions, it expects my compliance, and it calculates my more or less obedient behavior as a major investment in the sonic capital it exerts. My individual and personal, sensory situation *factually is* the capital of this economy. The market of sonic labor exploits exactly these substantial sensory situations and sensory sensibilities of the sonic workforce. The material working culture exploits the fields of location and apparatus, skills and habits, conflicts and heuristics, portfolio and presentations as inhabited by the individual sonic prosumer. The affective and caring work of these prosumers is even the main resource for an economy of sound—and their genuine ambition results in their imminent exhaustion that increases incessantly their exploitability. This is a form of generative labor that resembles almost a sonic capitalist's dream.

This dream of total exploitability expands into the material of individual, personal lives of sonic prosumers, promoting a fourfold, sensological forcefield of authoritarianism: *consumption, domestication, dispositive*, and *ubiquity.* Implied in this forcefield—resulting from the process of a mediated and globalized industrialization in recent decades—is a political system that is not only bound to an economic system: this new authoritarian system is identical on all levels with an economy of sensory exploitation, farming, and repression. The forms of sonic and sensory repression are successful as long as you or I enjoy living with them, working for them, contributing to and consuming them. As sonic consumers, we are claustrophobically enveloped in articulations, requests, and everyday practices of this pervasive sensology. You and I domesticate this sound culture and sound economy as we enjoy it, as we play with it, as we perform—maybe carnevalistically mocking or radically dysfunctional—those sonic patterns that we encounter. These sonic technologies of the self make our lives joyful and rich—and at the same time they become functioning modules in the economy of sound and in the panacoustic society. You and I, we enjoy how all our single moves are recorded and answered with sonic signals, we expect our every single action in the public sphere and in our personal spaces, in intimate relations to be recorded, to be exploited, to be calculated, and to be fed into the generative labor, developing new sonic artifacts and new sensory commodities providing future pleasures for us. The sensologies of surveillance do not represent a threat or a horrifying dystopia to us, we have embraced them, began to love and to crave them even, we digested their results, their apparatuses, their probes. We are an intrinsic function in their operations—and our functional, obedient, and well-trained listening practices are necessary for their operations. My work as a sound designer or a sonic prosumer contributes to this expanding surveillance state. The silencing dispositive is running smoothly and my belief in the optimization by corporeal and sensory training and refinement contributes every single day to the maintenance of this state of sonic capital. I am trained in easily ignoring the actual, the hurting, and long-lasting, damaging effects of the material sounds, the material apparatuses, the material relations to other sonic prosumers. If I fail to comply or to succeed I am trained in believing that this might all just be my personal fault, I need to train better, I need to optimize my time management, my sensory management, my media consumption, my desire to communicate or to relate to other sonic and sensory prosumers: I need to work harder, party harder, get harder in order to endure this lifestyle and the requests of the sensory workforce and the panacoustic community. Sure, some might have said, in earlier decades, that such a life resembled a total dystopia; but today, don't we enjoy this life, don't we indulge in the manifold pleasures it delivers to us, don't we simply profit from the advancement in globalized industrialization and connectivity, don't we enjoy the smooth refinement and the incredible subtleties of new commodities catered to our desires? This sensology is so delicious and so attractive and so desireable as it incessantly and ubiquitously cares for us. This is the situation of sound. This is the situation of the sonic capital.

Critical nanopolitics of sonic labor

What could be the contribution of nanopolitics to a critique of the sonic capital dominating sound design in the twenty-first century? Living with functional sounds entails a whole range of practices of listening, corporeally resonating, and contributing to and actually serving a sonic economy. Sonic labor is, however, not statically implemented in your or my body. It is a political and an economic concept; and as such it is subject to change. "It is worth recalling that what is currently called realistic was itself once 'impossible'" (Fisher 2009: 17). The core of sonic capitalism and of sonic labor is the very immanence of a situation of *prosumption*: a situation in which sonic patterns and sonic artifacts are experienced, invented, experimented with, refined, and finalized. This situation can be a starting point. This start is made possible, as soon as it is acknowledged that—again following from sonic materialism—it is the materiality of the generation, the propagation and reflection, the resonances as well as the sensed and audible effects of sound that contribute to the actual crafting of new sonic artifacts. Consumer and producer are no longer distinct categories to understand sonic labor by: this conflated situation is the actual core of sonic labor and at the same time its most vulnerable resource, its most easily attacked substance. This contemporary sensology can *not be* attacked on the fields of epistemology or economics, not on the fields of argument or logic, of historical responsibilities or political conventions; this sensology can, however, be attacked and questioned in its precise performativity, its actual movements and corporeal appearances, its material emanations and its very tangible and effectual presence. Sensologies as sensory ideologies of presence and tactics can most directly be attacked by interventionist tactics shifting the very present situations. Or, in other words: the repressive tolerance of sensological tactics can be repelled most effectively with situationist tactics disrupting exactly this newly established sensory performativities of repressive sensological tactics. To destroy a hegemonic sensology you might need foremost a disruptive sensology that dismantles sensologies as such—a truly paradoxical endeavor.

The paradoxical tactics needed here, are thus *nanopolitical tactics*. Sensologies in themselves are already *nanopolitics*: they are neither explicitely discussed as major political or economic goals, but they are entering everyday lives by way of all the microdecisions one makes, by way of sensory nanopreferences and routines, by way of inclinations and aversions you and I might tend to follow. These myriad of inclinations and aversions are openly guided and redirected under the forces of authoriarianism—consumption, dispositive, domestication, and ubiquity—and are easily enforced in a panacoustic society. Consequentially, there is not much upfront resistance even thinkable against this sensology and its nanopolitics. One might object in complex treatises or studies—maybe like this one—in vocal statements and openly hostile letters or public speeches directed toward responsible personnel; yet with every purchase in an online store, with every journey in a tracked public or individual transport vehicle, with every sign process enacted on a workstation, tablet, smartphone, or wearable device, you and I confirm and stabilize through our performativity this ruling sensology. You contradict your words with all your minuscule, innumerable microactions every single day. I contradict my words with all my minuscule, innumerable microactions every single day. Nanopolitics dig deeper and root deeper than the commonly reflected forms of politics (*cf.* the larger discussion in: Schulze 2018: 163–166): Whereas *macropolitics* represent the well-known forms of national governance through party politics, national institutions, and corporations ("politics"), the concept of *micropolitics* has been introduced in recent decades to describe the politics of local or communal organizations, companies, and associations

(Burns 1961/2; Mintzberg 1983; Neuberger 1995)—the area of nanopolitics is situated underneath this: nanopolitics govern and predetermine all the individual, sensory and corporeal constellations, situations and practices between body politics, perceptual politics, and pragmatics. Or, condensed in one research question as a resistance issue: "how to think politics with and through the body?" (Hansen, Plotegher, and Zechner 2013). Translated into the present questions of a cultural theory of sound design: *how to perform nanopolitical tactics in the realm of sonic labor and sonic capitalism?*

First of all, nanopolitical tactics are by definition *subversive*. They are by definition only possible while contributing to and being an intrinsic part of contemporary social, political, and economic institutions. Decisions of disconnecting, of leaving and *not* contributing in any way to the present panacoustic society for instance will only result in an unhindered development of exactly this feared, detested, and criticized societal construct as planned. Not engaging and not taking part in the institutions one intends to alter subversively would simply not be a choice. It would simply mean, not to execute any form of subversive counteraction but to bypass this option and to carry on as usual. This necessity is the genuine, inner contradiction of all subversive strategies for social and political change: to perform nanopolitics it is necessary to be a part of the societal construct in question and to learn and embody its versatile, experienced knowledge of economic, political, and social practices and processes. This leads directly to the second characteristic of nanopolitics: one needs to take part in the innermost center of a societal construct of ideological and false consciousness—just in order to potentially, at various points in the future, transform selectively or substantially this ideologically flawed, societal construct. But will one ever *really* be able to do so? Or will one just be spoiled by and assimilated into the ideological construct and never be able to actually perform a subversive action; in the end, then, all will have been a futile and useless illusion. This risky paradox of any subversive strategy is very present in nanopolitics. It poses the imminent danger of being so much invested in all the smooth procedures of present sensologies of surveillence, for instance, that a once subversive intention simply vanishes in the rapid vertigo and haze of everyday professional activities embedded in the silencing dispositive. If you actually hold the core beliefs of *sonic transcendence* or *audiopietism* (*cf.* the sections "The silencing dispositive" and "Functional listening" in Chapter 13) as true at one point then all subversive goals might already have been lost: the hungry sensological monster, this supermetamegainfrastructure has devoured you already, swallowed you whole. Thirdly, aside from these risks of subversion, there are very specific, concrete, and situated tactics for nanopolitical resistance. A hegemonic form of nanopolitics can be countered and questioned, expanded and transformed—unsurprisingly—by deviant nanopolitics. Some of these tactics have already appeared throughout this book (especially in the sections "Decolonizing sound design" and "The remediology of sonic labor" in Chapter 12)—and at this point in the final chapter I would like to resume some of their most operative and most easily applicable practices.

For sonic prosumers like you or me the situation of sonic labor is ubiquitous. This very situation of production, teamwork, use and misuse of technology, and inventing new concepts for sensotechnocultural practices is therefore the best contact point for nanopolitical tactics. The transformation and the alteration of these situations of generative labor, of their material constellations, time structure, spacing models as well as hierarchic conventions of communication and controlled processes of delivery are the arsenal at which nanopolitical counterattacks can be directed: one might change the venues for a workshop to another part of the city or the country, closer to the actual consumers of the client's products; one might arrange surprising and counterintuitive constellations of performers, musicians, and craftsmen to contribute; one might deny and disrupt the deadlines for presence, admission, and delivery that have been set by a production partner and are incompatible with a deeper sonic research; one might rely on specific open source or extremist sound libraries; one might establish another exchange routine between the team members that denies, disrupts, or even reverses verious kinds of existing hierarchies in the production team; and one might, last but not least, establish an identical standard honorarium for all members of the team—be they in a leading, a side, a supporting, or a background position. All of this might seem irrelevant and ridiculous, maybe even useless and superfluous, to care for the material appearances of individual situations—here, of designing sound—when the whole sensological system is not attacked fundamentally. But, actually, these nanopolitical transformations exchange playfully major constituents of the established dispositives of sonic labor: they might seem only superficially, only sensorially effective (venues, team building, feedback routines, billing) but they are surgically extracting and exchanging labor practices. If one would then decline any of these forms of resistance and counterattacks, one would then perform more of a form of resignation, capitulation, and focusing on mere decorative and more accidental aspects of everyday life. If a criticism of the missing fundamental critique in these nanopolitics were actually justified then sensological systems would be easy to change. Then, the gatekeepers of these systems—very often the clients and employers of sound designers—would be open to logical counterarguments, to a discursive exchange of ideas and concepts on the same level and both parties would share the same practices of discourse. A sensological institution operates though in a different continuum than the one of argument, logic, and conclusions. Consequentially, its gatekeepers need to consider approaches of critique and discourse as ridiculous, as wrong and superfluous, as only the maintenance of the installed sensological structure and the destructuring of any other structure is central to them. Everything else outside this major goal of destructuring and destroying any reasonable discourse open to critique seems negligible to then. Hence, the situations of actual experience and of actual sonic labor are also neglected, ignored, and denied existence. This is the contact point. This is the battleground.

Critical nanopolitics of sonic labor, therefore, need to transform foremost the actual, minuscule actions and the actual, situated impact of the silencing dispositive: propose a sound concept without any visual and written outline; let a sound concept be developed by the factual users and consumers, with the most erratic freedoms you can grant them; get unemployed users and lower paid employees of the client's company to decide about the sound design; explore at length, for days and weeks and months and years the actual listening situation where the sound design will be implemented; play erratically with the weirdest and most disrupting sound designs being played in this actual listening and usage situation; let understandable information about the precise political and economical background of this very sound design enter into the sound design you propose. A first step in such nanopolitics is to exclude the truisms and hyperstitions of audiopietism, and a second step would be to also exclude the silencing and the filtering of annoying and disturbing sources as the goal of sonic labor. As in the brief interventions mentioned just above, these nanopolitics do not aim at an untainted, self-serving sound design, but at a sound design that incorporates massive traces of the noise and the social and political situation into the design. When the two main truisms of contemporary sonic labor are being dismantled a quite different generativity of sound design might then emerge. As soon as the belief in a self-satisfied audiopietism and its silencing dispositive as core sensologies no longer exists then the actual personal, sensible, not yet predetermined interpersonal and societal aspects of caring and of affective labor can be increasingly stressed in sonic labor. The inner contradictions of sonic as free and affective labor are perfectly articulated by Tiziana Terranova:

> The increasingly blurred territory between production and consumption, work and cultural expression, however, does not signal the recomposition of the alienated Marxist worker. The Internet does not automatically turn every user into an active producer, and every worker into a creative subject. The process whereby production and consumption are reconfigured within the category of free labor signals the unfolding of a different (rather than completely new) logic of value, whose operations need careful analysis. (Terranova 2000: 35)

The actual situation of listening and sounding, of generating sound and crafting sound is therefore not magically reversed and altered; the whole material impact of situated listening and sounding is taken and transformed into one that "emphasizes how labor is not equivalent to employment" and stresses "how important free affective and cultural labor is to the media industry, old and new" (Terranova 2000: 46).

As soon as the massive workloads of affective and caring and cultural labor are accepted as an integral, core part in the process of sonic labor (and generative labor in general) then they might even get billed for and paid, and possibly aim at reorganizing and dismantling sonic labor in general. In the end the complex value of sonic labor would be critically and generatively translated into present exchange value formats—even in contemporary sensological societies—rendering them immediately incoherent and, subversively, utterly useless. This expansion of affective labor—of detailed and overly detailed work into all constituents of a contemporary work situation (e.g., material constellations, time structure, spacing models, hierarchic conventions of communication, controlled processes of delivery) in relation to all, even the most negligible sonic prosumers—could in the end of a long, subversive, and factually revolutionary process overstretch sonic labor. It might implode; it might tear up: it might simply dwindle away. In any of these cases, a transformation takes place that does not leave sonic labor as it was before. With an emphasis on *affective labor as part of sonic labor*, the silencing dispositive can be defused, the self-optimizing orders of audiopietism are rendered useless, and the whole militarized ideal of effective and undisturbed communication and execution of sequenced goals—or the notorious *milestones* in *strategic planning*—become meaningless. As no fraction of a societal construct is thoroughly independent from all others this might surely have effects on connected areas: they might then also transform, even if only slowly, they might also attract subversive forms of nanopolitical behavior. Along the lines of a more coherent theory of social, economical, and political transformation, of reform and revolution, this could represent one starting point of an accelerating, intensifying, amplifying domino effect. I am far from claiming that any transformation in the area of sound design will lead to a political transformation involving the whole transnational sequence of economies, lifestyles, and cultures. But every transformation of sonic labor will contribute a significant trace element to an accumulating desire and sense for a larger political and social transformation.

After some of the tactics of critical nanopolitics that I mentioned earlier have done their deed, quite different spaces are filled with different actors with quite different habits and goals that interact therein in quite different ways. Still, they might generate sonic artifacts—but they surely won't eject an endless line of functional sonic signals following the militarized scholastics of hunting signals. The rather limited and reductive signaling set of *orders, contact signals, informative signals, mood signals, signals with heightened symbolic meaning* (Karbusický 1986; *cf.* Chapter 9, sections "The semiotics of hunting signals" and "Performing sonic patterns") might then be abolished. Instead the decolonized, remediated, and regenerated sonic affordances provide the material to be crafted sonically. Critical nanopolitics resignify and reenact each step in sonic labor in a new and different and sensorily sensible, reflected, and regenerated mode. A new cultural transmission of new sounds and a renewed design culture would then begin. Redesign culture?

Generativity, remediation, regeneration

How can practices of remediation and regeneration provide the main source of generativity for sound design? To work in sound design and to invent, craft, and implement a sound design requires activities that are capable of establishing sounds in a sonic environment that can contribute to, escort, or even inspire activities. These activities can be limited to a restricted set of actions demanded from anyone listening—as in the case of *alarm signals*—but they can also be opened up to a wide range of unintended yet still implied activities. As soon as work in sonic labor can apply some of the various nanopolitical tactics proposed in the previous section, then the process of sonic labor is substantially altered. It can envision different goals and it can apply more strategies and practices than just to reach the milestone of presentable version material for a number of sonic signals and their usage instructions. The different goals are goals that are intrinsic to the process of crafting and designing. I would like in this section to resume discussing the transformations that can take place in the situation of sound design and in the process of sonic labor under three core concepts and major tactics of nanopolitics: *generativity, remediation*, and *regeneration*—with its accumulated effect of *decolonization*.

The first concept, *generativity* is also the defining approach at the core of generative labor. It refers to the activity of generating not only individual products out of given work assignments, sonic patterns, and production timelines but to the activity of genuinely inventing and refining fundamental ideas and concepts of how to craft a sound design, how to detect concealed sonic affordances, and how to actually craft new and maybe unheard sonic patterns. The whole process of sonic labor is part of this generative labor that moves this kind of labor closer to all kinds of artistic and performative activities. Generative labor invents and generates new concepts, new ideas, new constellations of connections. It is the labor of invention and of actual creation, may it use aleatoric, improvisational, or combinatoric strategies to invent (Schulze 2000) or may it delegate some of its invention and problem solving strategies, its heuristics (Schulze 2005) to remote repositories, mechanics, or mechanistically approached software as expert systems. To actually acknowledge this activity as the core of one's sonic labor can transform the situation of labor: suddenly, this labor is no longer about meeting deadlines, delivering products, being present at checkpoints and being granted admission to a work process—all of which would result from the controlling dispositive of *admission, presence*, and *documentation*. It is now more necessary to focus on the intrinsic motivation and the intricate generativity in working life, and to expand its quality and impact. Following generativity as a guideline for sonic labor it is necessary first to minimize most of the planning, organizing, legimitizing, ritualized controlling and silencing processes of admission, presence, and documentation; and secondly to maximize the time and the space, the situations, and the collective or the isolated collaborations on actual generativity. As soon as generativity can be maximized and can be reinstated as the actual core or sonic labor, then the character of employment (*cf.* Terranova 2000) fades away. Yet, today the established practices of coercing generative labor into the given routines and surveillance structures are still powerful, with the incessant extortion through threats at times being even dangerously life threatening (Hesmondhalgh and Baker 2011). Generative labor is still subject to the panacoustic society and its lines of control, of presence, admission, and documentation. The concept of *generativity* hence implies also the decolonizing practice discussed earlier as: *shifting the situation of sonic labor* from a commodified and formatted to a situated and corporeal one (*cf.* Chapter 12, section "The remediology of sonic labor"). A generative labor liberated from the aforementioned restraints might be able to contribute substantially to a decolonizing of sound design artifacts.

Remediation, proposed here as a second main tactic of nanopolitics, refers to the well-known media theory approach (Bolter and Grusin 2000; *cf.* Chapter 12, section "Remediating sonic patterns") combined here with the concept of transmission from mediology (*cf.* Chapter 11). Such a refined concept of remediation is complex enough to sufficiently interpret any advanced media transformation and will be applied here to sound design in particular. Remediation in the reading proposed here represents a tactical intervention of using, reusing, misusing, transforming, applying, reversing and thus *remediating* earlier developed media products such as—in our case—sonic signals, sonic patterns, sonic affordances, and sonic artifacts. This can apply to the reusage of various sound libraries in order to manipluate and incorporate them, but, more ambitiously, it applies to the effort of excavating maybe forgotten or neglected sonic patterns that can be reused in a very appropriate and helpful manner. In the tradition of Western composition and in many classical, modern, and vernacular musical cultures around the globe, such forms of reapplying, transforming, and simply *playing with* known tunes, variation structures, compositional patterns or structural choices for arrangements, or instrumentation in one composition are well known. In this field of practice they constitute factually a major foundation upon which new performance styles, new compositional aesthetics, hardware or instrumental innovations, new musical interpretations, and new works can be based. *Remediating music* is the basic task of instrumentalists, conductors, and composers. In the much newer profession of sound design, however, with its strong focus on licencing and on the harshly restricted use of proprietary material, the task of remediating suddenly appears to be subversive, provokingly new, strategically explosive, and innovative. In this field of tension sound designers are used to working nowadays, the remediation of existing sonic signals, sonic affordances, or sonic patterns represents a basic copyright infringement—and hence a blasphemic deed in the economy of sound design—therefore, sound designers constantly have to prove how innovative, new, and thus marketable their design is. This might sound like a side note but it actually affects the core of contemporary design. The corporate horror if confronted with remediated, reused, and repurposed sounds mirrors quite nicely the imaginary scarcity of sonic or even musical ideas; a truly strange if not inconceivable horror in times when most of the activities in coding, language operations, and media products are obviously only possible by reusing existing modules, repurposing existing formulaic structures, and remediating existing patterns. The illusion of radical invention, however, seems still to be rampant in design. Radical remediation, to the contrary, denies effectively the mere existence of radical invention and relies, hence, mainly on reusing, remodeling, and even reversing existing material. Also from this background stems the use of remediation for a decolonizing critique of existing artifacts (*cf.* Chapter 12), as it allows a thorough revision and reversal of contemporary practices. With advocating the already widely existing and highly prolific practices of remediation in sound design as a major and legitimate and legal practice that every single sound designer has to use all the time, the illusion of radical invention might disappear one day. At the same time this might draw, surprisingly so, even more interest to *actually* radically new and hitherto unknown sonic patterns, or even design patterns, that emerge anew from new areas of sonic labor. The concept of *remediation* hence also implies the decolonizing practice discussed earlier as: *shifting the sonic material* from a commodified and formatted to a situated and corporeal one (*cf.* Chapter 12, section "The remediology of sonic labor"). This second core concept might also contribute to decolonization as a substantial and thorough rethinking and reinventing of sound design artifacts—rejecting and subverting the established hierarchical media genres of colonization.

The third and final core concept of nanopolitics is *regeneration,* resulting from both generativity and remediation. Regeneration as a practice combines the generative flow of concepts, ideas, weird and old in order to reorganize, reformulate, and remake existing concepts, ideas, and artifacts. This practice actually provides the framework for generative labor in general; but in the form of *regenerative labor* this practice denies space and time to the common and colonized practices of exploiting and combusting of humanoid aliens. Instead of often exploitative generative routines in the area of sonic or sensory labor, this regenerative labor expands time and space with a careful respect for the needs and desires, the individual affordances and preferences of the sonic consumers themselves: a sustainable and a materially realistic form of labor respects the bodies, the sensibilities, the minds and lives, the sensory idiosyncrasies and the individual habits of those working in it. *Regeneration, thus, means that the framework of employment for the production of sonic artifacts is altered to a framework of intrinsic joy of the partaking sensory corpora undertaking sonic research in relation to existing artifacts, a practice of remediating generativity.* As abstract as this sentence might sound, it no longer grants leadership, guidance, and dominance to predetermined schedules, structures, budget decisions, or flowcharts but to the sensibilities of actual and personal, physical and idiosyncratic listening and sounding bodies, to the *sonic personae*. They are no longer reduced to sonic prosumers, but they are allowed to expand, they are encouraged to play out all the richness, the repleteness of their lives in exploring sonic affordances. Sonic domestication in performative excess is the key practice here: the sonic personae crafting, inventing, and remediating sonic patterns domesticize and accommodate themselves and a given listening and sounding situation—a multitude of listening and sounding personae—to a new but adequate sound design artifact (*cf.* Chapter 9, sections "Performing sonic pattern" and Chapter 10, section "Domesticating sounds"). It might well be that new sonic personae emerge from such processes, personae of humanoids that are no longer restrained by an economy of sound but find themselves embedded in regenerative everyday life practices:

> A sonic persona is made out of a sensory corpus struggling with changing auditory dispositives. In these struggles, one negotiates a viable persona. (Schulze 2018: 157)

Sonic personae therefore represent the state of one's struggle. A life with sounds is thus possible, desireable, and liveable for a humanoid like you or me. Hence, the concept of *regeneration* also implies the decolonizing practice discussed earlier as: *shifting the listening practice* from a commodified and formatted to a situated and corporeal one (*cf.* Chapter 12, section "The remediology of sonic labor"). Therefore, this third core concept might also contribute to a decolonization as a substantial and thorough rethinking and reinventing of sound design artifacts—rejecting and subverting the established hierarchical constellations of listening consumption of colonization.

The situation as interpreter

How might sound design change if the situated and corporeal experiential situation is established as its main interpreter? We listen to the sound you crafted. We gather in this fine studio complex in the basement, at the end of a long and weirdly labyrinthic aisle of this old building from the late nineteenth century, in an industrial area outside of the city. The finest machines for recording, postproduction, and mastering can be found gathered here. Most of the musicians and agency people, producers, and clients present in this room I have never met before. But we share the love for your work, we escorted your work process for this new sonic artifact produced for a major, international corporation. We signed nondisclosure agreements, for various aspects of your work, months ago, way before your actual work began. What we will hear today, on this very late afternoon of a hot summer day, will sound completely different than you and I and most of the other listeners invited to this premium presentation environment will have expected. I have no doubt about that.

The series of sound events you designed—the minuscule and subtle sonic signals you invented over time, the sonic patterns you imagined, tested, refined, tested again, then restructured, even further simplified, and again tested, then layered and tested again, revised again, and then finally implemented—all of this detailed work—widely scattered over several, long months, including travels to headquarters and franchises of this company, meetings with higher ranked, lower ranked and medium ranked personnel of this corporation, all these hours and days and weeks alone in the studio, at your laptop, in smaller meeting rooms, coffee shops, bars, trains, cars, planes together with selected members of this production team of yours—now gets confronted with an actual situation of listening. This situation of listening though consists of a specific selection of materials, objects, displays, smells, spatial enclosures, apparatuses, furniture, bodies, expectations, and corporeal listening habits that might just represent one, very particular situation of listening. This whole listening situation represents a strong enclosure and an even stronger framing for your sound design. This framing though might be very different in everyday use, for example, with all the manifold overlays and interruptions that are missing here and now: usually the sound design is just a minor source with people talking over it, with bodies moving, interacting, working with the device that emits those sounds, with musics overlayed, other conversations, other sounds of other devices. But the once carefully designed sonic artifacts are now out there, out here in these wild and entangled situations of usage. Still, you, the sound designer, hope the sonic affordances you discovered work more or less the way you expected and you hope also that the consumers of your sounds do actually assimilate them, appropriate, domesticate them more or less in a way that does not really effectively destroy them (as demonstrated in Chapter 7, section "Design abuse"). We sit now in this highly artificial listening environment and effectively we all realize that this listening situation lacks the laconic disturbances, the weird interruptions, also the neglect and disinterest that is so characteristic for any actual usage of sound designs. As much as you will try to foresee, to plan ahead, to include the actual usage situation in your design, the effective situation of actual use remains still the hardest decisive environment in which your inventions can be tested. The situation of use *is* the actual interpreter of your sonic artifact.

What might sound like a trivial truth for the field of design in general is not yet an accepted axiom for the more specific field of sound design. At least not in the sound design projects that constitute the empirical foundation of this book—ranging from designing the functional sounds of endoscopic machines for a surgery room, the game sounds of a newly developed casual game on various smartphone platforms, the sounds for a new media entertainment system up to the sounds for a large, international railway company. In almost all of these cases the sound designers had indeed a hard time of convincing their clients that what they heard in a listening studio or at their office computers is at best *physically* identical with what any user might hear—but not at all what they might experience *in situ*. From the client's side there is often a strong belief in sonic transcendence, in the strong overall-effect that music and sound has in general—no matter from which device, in what environment, for what listeners, at what time of the day, or in what cultural environment a sound event occurs. The material, mediated, acoustic, and performative requirements that allow us to perceive certain sonic patterns are easily ignored. This ignorance, however, damages the actual sound design. If one believes sound design could work independently from material constellations and physical requirements, then one believes in audiopietism and sonic transcendentalism. Effectively, one then does not intend to listen but just "to also have some sounds in here": a horrifying phrase sound designers have to hear a lot.

The meaning of sounds, their interpretation, and the understanding of sounds (addressed especially in Chapter 9, section "Rematerializing sonic signs") is however—also in the case of *designed* sounds—an effect of their individual, situated entanglement. Only the actual usage situation is the best testing ground for any sound design where the impact of sonic materialities can unfold. As long as a sound designer or a client ignores the major impact of materials on sound generation, sound propagation, and sonic consumption then the crafted sound design will surely more resemble a mock-up for some conceptual idea behind it than an actual fully developed design. Only as soon as your sound design is embedded into at least various exemplary listening situations with all their highly intricate and malleable constellations of materials, only then is it at least possible to listen to the sound design in the most appropriate listening situation. In pursueing this goal of testing sound designs in highly realistic and detailed listening situations one major lack of earlier processes of sonic labor can be countered: the common task of designing sound for material objects not yet known. In numerous cases of our field research and case studies, we encountered exactly this one major flaw: the sound designers were not given any major insight into the situation of usage in all its minuscule details of materials, software standards, audio technology standards used, the precise audience situation, or the relevant listening environment: What did the audience listen to before? What will the audience maybe listen to afterwards? The sonic artifact in the end might simply not work at all in these situations and their material conditions.

Following the insights of previous chapters on sonic materialism and sensologies, on the activities of sonic consumers and the work in sonic labor, on the situatedness of sonic affordances and the process of domesticating sonic patterns, the limits of functional listening and the power of the silencing dispositive, there is simply no way around one final insight: the quality of crafting sound—be it for narrowly functional purposes or for more experimental goals—is almost directly connected to the exploration of (again following Régis Debray here) the organized material of devices, architectures, and decorations and the material organization of actors, listeners, and passerbys (as elaborated in Chapter 11, section "Translating mediology"). Whereas both categories described in Debray's original mediology are foremost conceptualized for long-haul transmissions of cultural practices and apparatuses into an unknown future, precisely both these categories can also be applied to explore the situated and instantaneous situations of emitted sounds and sonic signals. Mediology's three analytical questions (introduced in Chapter 11, section "Three mediological questions") can therefore be used as guiding questions for a sound design process, as a kind of *sonic mediology: What are the material artifacts involved in this process of cultural transmission?* This first mediological question reminds a sonic prosumer to consider how even the least constituent of a usage situation provides sonic affordances. *What are the meanings realized by this process of cultural transmission?* The second mediological question supports a sonic prosumer in distinguishing the actualized meaning of a sonic pattern from the meaning maybe intended by a composer or designer—a necessary step to explore how this meaning effectively unfolds in a listening situation. *What are the actors and their allied and strategic goals of a cultural transmission?* The final and third mediological question reminds a sonic prosumer of the many humanoid aliens, sonic consumers and their sensory corpora involved and active when a sound design is implemented—with all the mainfold interests, needs, desires, goals, and moments of ignorance. By answering these three questions in the process of designing sound the crafted artifact might be more appropriately embedded in its material situation of listening and sounding. The previously proposed decolonizeng practices as well as the core tactics of nanopolitics are then possible to enact in the process of sound design (*cf.* the previous section on "Generativity, remediation, regeneration" and Chapter 12, section "The remediology of sonic labor"). Sonic signals might then indeed have appropriate effects.

The material situation of sounding and listening is the ground for a process of sonic labor. This is where sonic capital can be heard, where it extorts its sensologies, were the critical work of and decolonizing needs to take place, and where tactical nanopolitics can alter the culture of sound design. It's all in the practices of listening and sounding on every single day, in every single instant, in the peculiar constellation of sonic materials and sonic personae. Right now, you are entangled into them—again, still. A brief and rhythmically structured, rather low volume, and not alarmingly high-pitched sonic signal was just emitted by the tiny speaker in my mobile phone. I had to smile.

REFERENCES

Agamben, G. (2009), *What Is An Apparatus? and Other Essays,* trans. D. Kishik and S. Pedatella, Stanford, CA: Stanford University Press.

Alexander, C. (1977), *A Pattern Language. Towns, Buildings, Construction*, New York: Oxford University Press.

Amin, S. (1985), *Delinking. Towards a Polycentric World,* trans. M. Wolfers, London: Zed Books.

Ansari, A., D. Abdulla, E. Canli, M. Keshavarz, M. Kiem, P. Oliveira, L. Prado, and T. Schultz (2016), "Editorial Statement." *Decolonising Design*, June 27, 2016. Available online: www.decolonisingdesign.com/statements/2016/editorial/ (accessed November 1, 2018).

Appadurai, A., ed. ([1986] 1996), *The Social Life of Things. Commodities in Cultural Perspective*, Cambridge: Cambridge University Press.

Aranda, J., B.K. Wood, and A. Vidokle, eds. (2015), *The Internet Does Not Exist*, Berlin: Sternberg Press.

Attali, J. (1977), *Bruits: Essai sur l'économie politique de la musique,* Paris: Presses Universitaires de France.

Attali, J. (1985), *Noise: The Political Economy of Music*, Minneapolis: Minnesota University Press.

Augoyard, J.-F. and H. Torgue (2005), *Sonic Experience. A Guide to Everyday Sounds,* trans. A. McCartney and D. Paquette, Montreal: McGill-Queen's University Press.

Auinger, S. and B. Odland (2007), "Hearing Perspective (Think with Your Ears)," in C. Seiffarth and M. Sturm (Hg.) (eds.), *Sam Auinger. Katalog*, Wien: Folio.

Balzacq, T., ed. (2011), *Securitization Theory: How Security Problems Emerge and Dissolve*, Oxford: Routledge.

Barbrook, R. and A. Cameron (1996), "The Californian Ideology." *Science as Culture* 6 (1): 44–72.

Baudrillard, K. (1994), *Simulacra and Simulation,* trans. S.F. Glaser, Ann Arbor: University of Michigan Press.

Baudry, J.-L. (1970), "Effets idéologiques produits par l'appareil de base." *Cinéthique* (7–8): 1–8.

Baudry, J.-L. and A. Williams, (1974/5), "Ideological Effects of the Basic Cinematographic Apparatus." *Film Quarterly* 28 (2) (Winter): 39–47.

Benjamin, W. ([1940] 2005), *Concepts of History*, trans. Dennis Redmond. Available online: http://members.efn.org/~dredmond/Theses_on_History.html (accessed November 5, 2018).

Bijsterveld, K. (2008), *Mechanical Sound. Technology, Culture and Public Problems of Noise in the Twentieth Century*, Cambridge, MA: MIT Press.

Bijsterveld, K. (2018), *Sonic Skills. Listening for Knowledge in Science, Medicine and Engineering, 1920s-Present*, London: Palgrave Macmillan.

Bijsterveld, K. and E. Cleophas (2012), "Selling Sound: Testing, Designing, and Marketing Sound in the European Car Industry," in K. Bijsterveld and T. Pinch (eds.), *The Oxford Handbook of Sound Studies*, 102–124, : Oxford University Press.

Bijsterveld, K., E. Cleophas, S. Krebs, and G. Mom (2014), *Sound and Safe. A History of Listening behind the Wheel*, New York: Oxford University Press.

Blesser, B. and L.-R. Salter (2007), *Spaces Speak, Are You Listening? Experiencing Aural Architecture*, Cambridge, MA: MIT Press.

Bobin, J. and T. Waititi, dir. (2009), *Flight of the Concords*, Season 2, Episode 10, "Evicted." Aired March 22, 2009, on HBO.

Boever, A. De and W. Neidich, eds. (2013), *The Psychopathologies of Cognitive Capitalism: Part One*, Berlin: Archive Books.

Boltanksi, L. and E. Chiapello ([1999] 2005), *The New Spirit of Capitalism*, London: Verso Books.
Bolter, J.D. and R. Grusin (2000), *Remediation: Understanding New Media*, Cambridge, MA: MIT Press.
Bollier, D. (2014), *Think Like a Commoner. A Short Introduction to the Life of the Commons*, Gabriola Island, BC: New Society Publishers.
Bonz, J. (2015), *Alltagsklänge. Einsätze einer Kulturanthropologie des Hörens*, Wiesbaden: Springer VS.
Bronner, K. (2004), *Audio-Branding: Akustische Markenkommunikation als Strategie der Markenführung*, Munich: Grin Verlag.
Bronner, K. and R. Hirt, eds. (2009), *Audio Branding. Brands, Sound and Communication*, Baden-Baden: Nomos Edition Fischer.
Bronner, K. and R. Hirt, eds. (2016), *Audio-Branding: Entwicklung, Anwendung, Wirkung akustischer Identitäten in Werbung, Medien und Gesellschaft—3. Auflage*, Baden-Baden: Nomos Edition Fischer.
Burns, T. (1961/2), "Micropolitics: Mechanism of Institutional Change." *Administrative Science Quarterly* 6 (3): 257–281.
Buzan, B., O.O. Wæver, and J. de Wilde (1998), *Security: A New Framework for Analysis*, Boulder, CO: Lynne Rienner.
Cale, J. and B. Eno (1990), *Wrong Way Up,* London: Opal Records.
Certeau, M. de (1984), *The Practice of Everyday Life*, Berkeley: University of California Press.
Chandler, D. (2007), *Semiotics: The Basics,* 2nd ed., London: Routledge Publishers.
Chion, M. (1990), *L'Audiovision,* Paris: Editions Nathan.
Chion, M. (1994), *Audio-vision: Sound on Screen,* ed. and trans. C. Gorbman, with a foreword by W. Murch, New York: Columbia University Press.
Coleman, J. (2007), "Aurality," in P. Strohm (ed.), *Middle English*, Oxford Twenty-First Century Approaches to Literature Series, 68–85, Oxford: Oxford University Press.
Cox, C. (2011), "Beyond Representation and Signification: Toward a Sonic Materialism." *Journal of Visual Culture* 10 (2): 145–161.
Cox, C. (2018), *Sonic Flux. Sound, Art, and Metaphysics*, Chicago: University of Chicago Press.
Crary, J. (2014), *24/7: Late Capitalism and the Ends of Sleep*, New York: Verso Books.
Cross, N. (2007), *Designerly Ways of Knowing*, London: Springer.
Debray, R. (1981), *Teachers Writers Celebrities: The Intellectuals of Modern France*, trans. D. Macey, New York: Verso Books.
Debray, R. (1992), *Vie et mort de l'image: une histoire du regard en Occident*, Paris: Gallimard.
Debray, R. (1996), *Media Manifestos: On the Technological Transmission of Cultural Forms,* trans. E. Rauth, New York: Verso Books.
Debray, R. (1997), *Transmettre*, Paris: Odile Jacob.
Debray, R. (2000), *Transmitting Culture*, trans. E. Rauth, New York: Columbia University Press.
DeNora, T. (2000), *Music in Everyday Life.* Cambridge: Cambridge University Press.
Erlmann, V. (2010), *Reason and Resonance: A History of Modern Aurality*, New York: Zone Books.
Eshun, K. (1998), *More Brilliant Than The Sun. Adventures in Sonic Fiction*, London: Quartet Books.
Fisher, M. (2009), *Capitalist Realism: Is There No Alternative?* Winchester: Zero Books.
Fisher, M. (2014), *Ghosts of My Life: Writings on Depression, Hauntology and Lost Futures*, Winchester: Zero Books.
Florida, R. (2002), *The Rise of the Creative Class*, New York: Perseus Book Group.
Forsey, M.G. (2010), "Ethnography as Participant Listening." *Ethnography* 11 (4): 558–572.
Foucault, M. (1995), *Discipline and Punish: The Birth of the Prison*, New York: Vintage Books.
Foucault, M. (1980), *Power/Knowledge: Selected Interviews and Other Writings 1972–1977,* ed. C. Gordon, trans. C. Gordon, L. Marshall, J. Mepham, and K. Soper, New York: Pantheon Books.
Flückiger, B. (2001), *Sound Design: Die virtuelle Klangwelt des Films*, Marburg: Schüren Verlag.
Flückiger, B. (2009), "Sound Effects: Strategies for Sound Effects in Film," in G. Harper, R. Doughty, and J. Eisentraut (eds.), *Sound and Music in Film and Visual Media: A Critical Overview*, 151–179, New York: Continuum Publishers.

Gaver, W. (1991), "Technology Affordances," in S.P. Robertson, G.M. Olson, and J.S. Olson (eds.), *Proceedings of the SIGCHI Conference on Human Factors in Computing Systems*, 79–84, New York: ACM Press.
Gendlin, E.T. (1992), "The Wider Role of Bodily Sense in Thought and Language," in Maxine Sheets-Johnstone (ed.), *Giving the Body its Due*, 192–207, Albany: State University of New York Press.
Gibson, J.J. (1977), *The Ecological Approach to Visual Perception*, London: Routledge.
Gillis, J., ed. (1998), *The Militarization of the Western World*, New Brunswick, NJ: Rutgers University Press.
Goldsmith, K. (2016), *Wasting Time on the Internet*, New York: Harper Perennial.
Goodman, S. (2009), *Sonic Warfare: Sound, Affect, and the Ecology of Fear*, Cambridge, MA: MIT Press.
Gopinath, S. (2013), *Ringtone Dialectics: Economy and Cultural Form*, Cambridge, MA: MIT Press.
Großmann, R. (2008), "Verschlafener Medienwandel. Das Dispositiv als musikwissenschaftliches Theoriemodell." *Positionen—Beiträge zur Neuen Musik* (74): 6–9.
Groves, J. (2011), *comMUSICation: From Pavlov's Dog to Sound Branding*, Oxford: Oak Tree Press.
Haider, S. (2017), "The Darkness at the End of the Tunnel: Artificial Intelligence and Neoreaction." *Viewpoint Magazine*, March 28, 2017. Available online: www.viewpointmag.com/2017/03/28/the-darkness-at-the-end-of-the-tunnel-artificial-intelligence-and-neoreaction/ (accessed November 1, 2018).
Hall, S. (1973), Encoding and Decoding in the Television Discourse (Occasional Paper, Media Series: SP No. 7, 1973), Birmingham: Centre for Contemporary Cultural Studies.
Hansen, B. R., P. Plotegher, and M. Zechner (2013), *Nanopolitics Handbook: The Nanopolitics Group*, Wivenhoe, UK: Minor Compositions.
Hardt, M. (1999), "Affective Labor." *boundary* 2, 26 (2) (Summer): 89–100.
Harris, M. (1996), *Proclaim Jubilee! A Spirituality for the Twenty-first Century*, Louisville, KY: Westminster John Knox Press.
Henrich, J., S.J. Heine, and A. Norenzayan (2010), "The Weirdest People in the World?" *Behavioral And Brain Sciences* 33 (2–3): 61–135.
Henriques, J. (2011), *Sonic Bodies: Reggae Sound Systems, Performance Techniques, and Ways of Knowing*, New York: Continuum.
Hesmondhalgh, D. and S. Baker (2011), *Creative Labour: Media Work in Three Cultural Industries*, London: Routledge.
Hobsbawm, E. and T. Ranger, eds. (1983), *The Invention of Tradition*, Cambridge: Cambridge University Press.
Hogan, M. (1998), *A Cross of Iron: Harry S. Truman and the Origins of the National Security State*, Cambridge: Cambridge University Press.
Howes, D. (2005), *Empire of the Senses: The Sensual Culture Reader*, Oxford: Berg Publishers.
Hoy, M.B. (2018), "Alexa, Siri, Cortana, and More: An Introduction to Voice Assistants." *Medical Reference Services Quarterly* 37 (1): 81–88.
Ismaiel-Wendt, J. (2011), *Tracks'n'treks: populäre Musik und postkoloniale Analyse*, Münster: UNRAST-Verlag.
Ismaiel-Wendt, J. (2013), "Track Studies: Popular Music and Postcolonial Analysis," in J. Gohrisch and E. Grünkemeier (eds.), *Postcolonial Studies Across the Disciplines*, 89–108, Amsterdam: Editions Rodopi B.V..
Jackson, D. (2004), *Sonic Branding: An Essential Guide to the Art and Science of Sonic Branding*, Basingstoke: Palgrave Macmillan.
James, W. (1912), *Essays in Radical Empiricism*, New York: Longman Green and Co.
Jullien, F. (2016), *Il n'y a pas d'identité culturelle: mais nous défendons les ressources d'une culture*, Paris: L'Herne.
Karbusický, V. (1986), "Jagdsignale als Zeichensystem." *Zeitschrift für Semiotik* 8 (3): 277–281.
Kassabian, A. (2013), *Ubiquitous Listening: Affect, Attention, and Distributed Subjectivity*, Berkeley: University of California Press.

Kassabian, A. (2015), "Ubiquitous Musics: Technology, Listening and Subjectivity," in A. Bennett and S. Waxman (eds.), *The SAGE Handbook of Popular Music*, 549–562, London: SAGE Publications.
Kauls, I., I. Grunberg, and M.A. Stern, eds. (1999), *Global Public Goods. Published for the United Nations Development Program*, New York: Oxford University Press.
Kinnamon, L. (2016), "Attention Under Repair: Asceticism from Self-care to Care of the Self." Special issue of *Women & Performance: A Journal of Feminist Theory* 26 (2–3): 184–196.
Krajina, Z., S. Moores, and D. Morley (2014), "Non-media-centric Media Studies: A Cross-Generational Conversation." *European Journal of Cultural Studies* 17 (6): 682–700.
Kramer, G., B. Walker, T. Bonebright, P. Cook, J. Flowers, N. Miner, and N. Neuhoff (1997), *Sonification Report: Status of the Field and Research Agenda*. ICAD. Available online: www.icad.org/websiteV2.0/References/nsf.html (accessed November 1, 2018).
LaBelle, B. (2010), *Acoustic Territories: Sound Culture and Everyday Life*, New York: Bloomsbury Publishing.
Lau, K.J. (2000), *New Age Capitalism: Making Money East of Eden*, Philadelphia: University of Pennsylvania Press.
Laurel, B. (2003), *Design Research: Methods and Perspectives*, Cambridge, MA: MIT Press.
Law, R. and M. Wark (2013), *W.A.N.T.—Weaponized Adorables Negotiation Tactics*, New York: Kickstarter.
Lawson, B. (2006), *How Designers Think*, Oxford: Architectural Press/Elsevier.
Lippe, R.Z. ([1987] 2000), *Sinnenbewußtsein: Grundlegung einer anthropologischen Ästhetik*, Reinbek bei Hamburg: Rowohlt (2nd ed. in 2 vols: Baltmannsweiler Schneider-Verlag, 2000).
Livingstone, S., P. Lunt, and L. Miller (2007), "Citizens, Consumers and the Citizen-consumer: Articulating the Citizen Interest in Media and Communications Regulation." *DISCOURSE & COMMUNICATION* 1 (1) (February): 63–89.
Lynne, J. (2015), "Reading the Numbers of Stop-and-Frisk." *Hyperallergic*, Brooklyn, NY: Hyperallergic Media Inc., September 15, 2015. Available online: https://hyperallergic.com/237694/reading-the-numbers-of-stop-and-frisk/ (accessed November 1, 2018).
Maier, C.J., M. Schneider, and H. Schulze (2015), "Situative Signals in Sonic Conflicts. Elements of a Sound Design Theory," in M. Bull and L. Back (eds.), *The Auditory Culture Reader*, 2nd ed., 119–133, London: Bloomsbury Publishers.
Maier, C.J. (2016), "The Sound of Skateboarding: Aspects of a Transcultural Anthropology of Sound." *The Senses and Society* 11 (1): 24–35.
Maras, S. (2008), "On Transmission: A Metamethodological Analysis (after Régis Debray)." *Fibreculture* 6 (12). Available online: http://twelve.fibreculturejournal.org/fcj-080-on-transmission-a-metamethodological-analysis-after-regis-debray/ (accessed November 1, 2018).
Mareis, C., M. Held, and G. Joost, eds. (2013), *Wer gestaltet die Gestaltung? Praxis, Theorie und Geschichte des partizipatorischen Designs*, Bielefeld: Transcript Verlag.
Mareis, C., G. Joost, and K. Kimpel, eds. (2010), *Entwerfen – Wissen – Produzieren. Designforschung im Anwendungskontext*, Bielefeld: Transcript Verlag.
Mau, S. (2017), *Das metrische Wir: Über die Quantifizierung des Sozialen*, Berlin: Suhrkamp Verlag.
McEnaney, L. (2000), *Civil Defence Begins at Home: Militarization Meets Everyday Life in the Fifties*, Princeton, NJ: Princeton University Press.
Mignolo, W.D. (2007), "Delinking: The Logic of Coloniality, and the Grammar of De-Coloniality." *Cultural Studies* 21 (2–3) (March–May): 449–514.
Mihm, A. (2016), "Baby Monitor: Parental Listening and the Organization of Domestic Space," in J. Gerrit Papenburg and H. Schulze (eds.), *Sound as Popular Culture: A Research Companion*, 215–224, Cambridge, MA: MIT Press.
Mintzberg, H. (1983), *Power in and Around Organizations*, Englewood Cliffs, NJ: Prentice-Hall.
Monelle, R. (2006), *The Musical Topic: Hunt, Military and Pastoral*, Bloomington, IN: Indiana University Press.
Montgomery, J. (2005), "Beware 'The Creative Class': Creativity and Wealth Creation Revisited." *Local Economy* 20 (4): 337–343.

Moore, J.W. (2014a), "The Capitalocene, Part I: On the Nature & Origins of Our Ecological Crisis," unpublished paper, Department of Sociology, Binghamton University.
Moore, J.W. (2014b), "The Capitalocene, Part II: Abstract Social Nature and the Limits to Capital," unpublished paper, Department of Sociology, Binghamton University.
Morley, D. (2009), "For a Materialist, Non-media-centric Media Studies." *Television & New Media* 10 (1): 114–116.
"Musikalische Hupen." (1911), *Der Kunstwart* 25 (3): 209–210.
Nancy, J.-L. (2008), *Corpus,* trans. R.A. Rand, New York: Fordham University Press.
Nardi, C. (2016), "Critical Listening," in J. Gerrit Papenburg and H. Schulze (eds.), *Sound as Popular Culture: A Research Companion*, 395–402, Cambridge, MA: MIT Press.
Nattiez, J.-J. (1976), *Fondements d'une sémiologie de la musique*. Collection Esthétique. Paris: Union générale d'éditions.
Nattiez, J.-J. (1987), *Musicologie générale et sémiologie*. Collection Musique/Passé/Présent. Paris: C. Bourgois.
Neidich, W., ed. (2014), *The Psychopathologies of Cognitive Capitalism: Part Two*, Berlin: Archive Books.
Neidich, W., ed. (2017a), *The Psychopathologies of Cognitive Capitalism: Part Three*, Berlin: Archive Books.
Neidich, W. (2017b), *Neuromacht. Kunst im Zeitalter des kognitiven Kapitalismus. Übersetzt von Jan Georg Tabor*, Berlin: Merve Verlag.
Neuberger, O. (1995), *Führen und führen lassen,* Stuttgart: Lucius & Lucius.
Obadike, M. and K. Obadike (2015), *Numbers Station [Furtive Movements]*, New York: Ryan Lee Gallery.
Ochoa Gautier, A.M. (2014), *Aurality: Listening and Knowledge in Nineteenth-Century Colombia*, Durham, NC: Duke University Press.
Oliveira, P. (2017), "Decolonizing the Earview opf Design: Listening Anxieties and the Apparatus of Auditory Governance," PhD diss., University of the Arts Berlin.
Oliveros, P. (2005), *Deep Listening: A Composer's Sound Practice,* New York: iUniverse, Inc.
Papenburg, J.G. and H. Schulze (2016), *Sound as Popular Culture. A Research Companion*, Cambridge, MA: MIT Press.
Peirce, C.S. (1998), *The Essential Peirce,* vol. 2., ed. Peirce Edition Project, Bloomington: Indiana University Press.
Perniola, M. (1991), *Del Sentire*, Turin: Einaudi.
Pink, S. (2009), *Doing Sensory Ethnography,* London: SAGE Publishing.
Pink, S. (2012), *Situating Everyday Life,* London: SAGE Publishing.
Plattner, H., C. Meinel, and L. Leifer, eds. (2010), *Design Thinking: Understand, Improve, Apply*, Berlin: Springer.
Posner, R. (2004), "Basic Tasks of Cultural Semiotics," in G. Withalm and J. Wallmannsberger (eds.), *Signs of Power – Power of Signs. Essays in Honor of Jeff Bernard,* 56–89, Vienna: INST-Verlag.
Power, N. (2014), "Soft Coercion, the City and the Recorded Female Voice," in M. Gandy and B.J. Nilsen, *The Acoustic City,* 23–26, Berlin: JOVIS Verlag GmbH.
Preciado, B. (2013), *Testo Junkie: Sex, Drugs, and Biopolitics in the Pharmacopornographic Era,* New York: Feminist Press.
Preuss, D. (1980), "Signalmusik," PhD thesis, Technische Universität Berlin.
Prinz, S. (2013), "Alltagskultur als Designkultur: Formen des Gebrauchs. Über die alltägliche Ordnung der Dinge," in Y. Milev (ed.), *Designkulturen: Der erweiterte Designbegriff im Entwurfsfeld der Kulturwissenschaft*, 33–44, Munich: Wilhelm Fink Verlag.
Publio Delgado (2016), "Dubai was lit—Harmonizator." *YouTube*, July 9, 2016. Available online: www.youtube.com/watch?v=K00WGRzQlII (accessed November 1, 2018).
Pysiewicz, A. (2009), *Klangkontexte zwischen Ökologie und Branding,* Aachen: Shaker Verlag.
Rice, T. (2012), "Sounding Bodies: Medical Students and the Acquisition of Stethoscopic Perspectives," in T. Pinch and K. Bijsterveld (eds.), *The Oxford Handbook of Sound Studies,* 298–319, New York: Oxford University Press.

Roland, A. (2001), *The Military-Industrial Complex*, Washington, DC: American Historical Association.
Sachse, P. and A. Specker (1999), *Design Thinking: Analyse und Unterstützung konstruktiver Entwurfstätigkeiten*, Zürich: vdf ETH.
Schaeffer, P. (1966), *Traité des objets musicaux*, Paris: Éditions Du Seuil.
Schafer, R.M. ([1977] 1994), *The Soundscape. Our Sonic Environment and the Tuning of the World*, Rochester, VT: Destiny Books.
Schneier, B. (2003), *Beyond Fear: Thinking Sensibly about Security in an Uncertain World*, New York: Copernicus Books.
Schön, D. (1983), *The Reflective Practitioner: How Professionals Think In Action*, New York: Basic Books.
Schoon, A. (2012), "Unmerkliche Eindringlinge. Versuch über akustische Kontrolle," in A. Schoon and A. Volmar (eds.), *Das geschulte Ohr. Eine Kulturgeschichte der Sonifikation*, 285–297, Bielefeld: Transcript Verlag.
Schramm, W., ed. (1955), *The Process and Effects of Mass Communication*, Urbana: University of Illinois Press.
Schrimshaw, W. (2017), *Immanence and Immersion. On the Acoustic Condition in Contemporary Art*, New York: Bloomsbury Publishing.
Schulze, H. (2000), *Das aleatorische Spiel. Erkundung und Anwendung der nichtintentionalen Werkgenese im 20. Jahrhundert – Theorie der Werkgenese, Bd. 1*, Munich: Wilhelm Fink Verlag.
Schulze, H. (2005), *Heuristik. Theorie der intentionalen Werkgenese — Theorie der Werkgenese, Bd. 2*, Bielefeld: Transcript Verlag.
Schulze, H. (2007), "Die Audiopietisten. Eine Polemik." *Kultur und Gespenster* 2: s. 122–129.
Schulze, H. (2016a), "Das Auditive Dispositiv. Apparatisierung und Deapparatisierung des Hörens," in B. Ochsner and R. Stock (eds.), *senseAbility. Mediale Praktiken des Sehens und Hörens*, 235–255, Bielefeld: transcript verlag.
Schulze, H. (2016b), "Resistance and Resonance. A Political Anthropology of Sound." *The Senses and Society* 11 (1): 68–81.
Schulze, H. (2018), *The Sonic Persona: An Anthropology of Sound*, New York: Bloomsbury Press.
Serres, M. (2008), *The Five Senses: A Philosophy of Mingled Bodies*, trans. M. Sankey and P. Cowley, New York: Continuum Publishers.
Shannon, C. (1948), "A Mathematical Theory of Communication." *Bell System Technical Journal* 27 (July and October): 379–423, 623–656.
Silverstone, R. (1994), *Television and Everyday Life*, London: Routledge Publishing.
Simon, H.A. (1969), *The Sciences of the Artificial*, Cambridge, MA: MIT Press.
Smalls, C. (1998), *Musicking: The Meanings of Performing and Listening*, Middletown, CT: Wesleyan University Press.
SomethingUnreal (2007), "Music using ONLY sounds from Windows XP and 98!." *YouTube*, August 8, 2007. Available online: www.youtube.com/watch?v=dsU3B0W3TMs (accessed November 1, 2018).
Spehr, G. (2008), "Funktionale Klänge. Mehr als ein Ping," in H. Schulze (ed.), *Sound Studies. Traditionen – Methoden – Desiderate. Eine Einführung*, 185–208, Bielefeld: Transcript Verlag.
Spehr, G. (ed.) (2009), *Funktionale Klänge. Hörbare Daten, klingende Geräte und gestaltete Hörerfahrungen*, Bielefeld: Transcript Verlag.
Spoiden, S. (2007), *Régis Debray et la médiologie. Études réunies*, Amsterdam, Rodopi.
Sterne, J. (2003a), *The Audible Past: The Cultural Origins of Sound Reproduction*, Durham, NC: Duke University Press.
Sterne, J. (2003b), *The Sound Studies Reader*, London: Routledge Publishing.
Sterne, J. (2011), "The Theology of Sound: A Critique of Orality." *Canadian Journal of Communication* 36 (2): 207–225.
Sterne, J. (2012), *MP3 – The Meaning of a Format*, Durham, NC: Duke University Press.
Steiner, P. (2014), *Sound Branding: Grundlagen akustischer Markenführung*, Wiesbaden: Springer VS.
Stoever, J.L. (2016), *The Sonic Color Line: Race and the Cultural Politics of Listening*, New York: New York University Press.

Supper, A. (2012), "Lobbying for the Ear: The Public Fascination with and Academic Legitimacy of the Sonification of Scientific Data," PhD diss., Maastricht University.
Szendy, P. (2016), *All Ears: The Aesthetics of Espionage*, trans. R. Végső, New York: Fordham University Press.
Szendy, P. (2018), *Plus d'oreilles (vers une panacoustique généralisée)*, Paris: Philharmonie de Paris.
Szepanski, A. (2014a), *Kapitalisierung – Bd. 1: Marx' Non-Ökonomie*, Hamburg: Laika Verlag.
Szepanski, A. (2014b), *Kapitalisierung – Bd. 2: Non-Ökonomie des gegenwärtigen Kapitalismus*, Hamburg: Laika Verlag.
Tagg, P. (2012), *Music's Meanings*, New York: The Mass Media Music Scholars' Press.
Terranova, T. (2000), *Free Labor: Producing Culture for the Digital Economy*, Durham, NC: Duke University Press.
Thompson, E. (2002), *Soundscapes of Modernity: Architectural Acoustics and the Culture of Listening in America, 1900–1933*, Cambridge, MA: MIT Press.
Thompson, M. and I. Biddle (2013), *Sound, Music, Affect Theorizing Sonic Experience*, New York: Bloomsbury Publishing.
Toffler, A. (1970), *Future Shock*, New York: Random House.
Treasure, J. (2009), "The 4 Ways Sound Affects Us." *YouTube. TED Global*, October 16, 2009. Available online: www.youtube.com/watch?v=rRepnhXq33s (accessed November 1, 2018).
Treasure, J. (2010), *Sound Business: How to Use Sound to Grow Profits and Brand Value*, Oxford: Management Books 2000 Ltd.
Truax, B. (1984), *Acoustic Communication*, Norwood, NJ: Ablex Publishing Corporation.
Ultra-red (2014), *Five Protocols For Organized Listening*, Cologne: Buchhandlung Walter König.
Vincenti, W. (1990), *What Engineers Know and How They Know It: Analytical Studies from Aeronautical History*, Baltimore: Johns Hopkins University Press.
Wark, M. (2012), *Telesthesia. Communication, Culture, and Class*, Malden, MA: Polity Press.
Wark, M. (2014), "Birth of Thanaticism." *Public Seminar*, New York: The New School for Social Research. October 3, 2014. Available online: www.publicseminar.org/2014/04/birth-of-thanaticism/ (accessed November 1, 2018).
Warnes, K. (2015), "Electronic Route Guidance Systems," in M. Garrett (ed.), *Encyclopedia of Transportation Social Science and Policy*, London: SAGE Publishing.
Weiss, P.P. (2015), *Wenn Design die Materie verlässt: Sound. Das Design der Emotionen, der Imagination und der Lebendigkeit*, Norderstedt: Books on Demand.
Wulf, C., ed. (1997) *Vom Menschen. Handbuch Historische Anthropologie*, Weinheim: Beltz Verlag.
Wulf, C. (2013), *Anthropology. A Continental Perspective*, Chicago: Chicago University Press.
Zuboff, S. (2015), "Big Other: Surveillance Capitalism and the Prospects of an Information Civilization." *Journal of Information Technology* 30 (1): 75–89.
Zuboff, S. (2017), *Master Or Slave?: The Fight for the Soul of Our Information Civilization*, London: Public Affairs.

INDEX

www.ingramcontent.com/pod-product-compliance
Ingram Content Group UK Ltd.
Pitfield, Milton Keynes, MK11 3LW, UK
UKHW050945290926
15470UKWH00005B/1654

9 781501 330223